Compendium of Objective Soil Science

Compendium of

Objective Soil Science

Editors

R.K. Kaleeswari
M.R. Latha
R. Rajeswari

SATISH SERIAL PUBLISHING HOUSE

403, Express Tower, Commercial Complex, Azadpur, Delhi-110033 (India)
Phone : 011-27672852, Fax : 91-11-27672046
E-mail : info@satishserial.com, hkjain1975@yahoo.com
Website : www.satishserial.com

Published by :

SATISH SERIAL PUBLISHING HOUSE

403, Express Tower, Commercial Complex, Azadpur, Delhi-110033 (INDIA)
Phone : 011-27672852 Fax : 91-11-27672046
E-mail : info@satishserial.com, hkjain1975@yahoo.com

Reprint, 2020

ISBN 978-93-81226-59-9

Composed, Designed & Printed in India

Preface

Soil Science is an extensive science encompassing knowledge of multivarious subjects. This creates a demanding situation while preparing for the examinations in Soil Science. Basic understanding of terminology is essential for complete understanding of the subject. Recently several changes have been brought in the examination patterns and objective knowledge testing has been included for a need preliminary screening get the frequently asked questions and concept at one place. The present book is a new outlook in objectives of Soil Science to provide an insight into possible questions for different competitive exams.

This book will be of immense help for the students for several competitive examinations such as ARS, IARI entrance examination, Civil services and entrance exam for Agricultural Universities. Authors acknowledge people and personalities who helped directly and indirectly in shaping the book in its present form. Model answers have been prepared from consultation and references of several standard sources in the subject. Authors acknowledge the sources from which answers have been drafted and in no way diminish their importance by taking away any credit due to them.

Efforts have been taken to proof read the book before printing. As this is the first edition of the book, authors will be ever grateful to the readers for giving constructive comments and criticisms.

The authors wishes its readers a grand success in examinations, illuminated career and prosperity in life.

Authors

Contents ?

Chapter-1

SOIL GENESIS AND CLASSIFICATION

Q.1. Who is the founder of modern Pedology?

Ans. V.V. Dokuchaiev

Q.2. What are the three spheres of earth?

Ans. Lithosphere, Hydrosphere and Atmosphere

Q.3. Which are the two abundant elements in earth crust?

Ans. Oxygen (46.6%) and Silicon (27.7%)

Q.4. Give examples of primary minerals.

Ans. Quartz, Biotite, Muscovite

Q.5. Give examples of secondary minerals.

Ans. Silicate clays, hydrous oxide, clays of Iron and Aluminium

Q.6. What are lithophile elements?

Ans. Elements which ionize readily or form stable oxy anions Eg. Fe, Mn, Al, Ca, Mg, Si

Q.7. What are chalcophile elements?

Ans. Elements which tend to form covalent bonds with sulphide Eg. Cu, Zn, Pb

Q.8. What are siderophile elements?

Ans. Elements which readily form metallic bonds Eg. Fe, Co, Ni

Q.9. What are atmosphile elements?

Ans. Elements which tend to remain as atmospheric gases Eg. N, He, Ar, Kr, Xe

Q.10. What are biophile elements?

Ans. Elements which tend to be associated with living organisms Eg. C, H, O, N, P, S

Q.11. Which is least resistant mineral for weathering?

Ans. Gypsum

Q.12. What is abrasion?

Ans. The physical weathering of a rock surface by running water, glaciers or wind laden with fine particles.

Q.13. What is index mineral?

Ans. Mineral that is resistant to weathering, Eg. Zircon, Garnet

Q.14. Which is easily weatherable mineral?

Ans. Olivine

Q.15. What is Petrology?

Ans. Study of science of rocks

Q.16. What is Petrography?

Ans. Study of description of rocks

Q.17. What is Petro genesis?

Ans. Study of origin of rocks

Q.18. What are stratified rocks?

Ans. Sedimentary rocks

Q.19. What is one angstrom (l°A) equal to ?

Ans. 0.1 nm or 10^{-10}m

Q.20. Give examples for orthosilicates.

Ans. Olivine, Pyroxenes, Amphiboles

Q.21. Give examples for phyllosilicates.

Ans. Biotite, Muscovite

Q.22. Give examples for Tectosilicates.

Ans. Feldspar, Quartz

Q.23. Give the name of calcium bearing feldspar.

Ans. Anorthite

Q.24. Give the name of sodium bearing feldspar.

Ans. Albite

Q.25. What are the types of potassium feldspars?

Ans. Orthoclase, Microcline

Q.26. What is the fundamental building block of silicate minerals in earth's crust?

Ans. Silica Tetrahedron

Q.27. What are the minerals formed due to weathering of primary minerals?

Ans. Secondary minerals Eg. Serpentine, Clay minerals

Q.28. What are accessory minerals?

Ans. Minerals occurring in very small quantities in a rock Eg. Magnetite, Pyrite

Q.29. Give examples for amorphous minerals.

Ans. Allophane, Chalcodony.

Q.30. Which minerals form the main part of sand, silt and clay fractions of soil?

Ans. Primary minerals - silt, sand
Secondary minerals - clay

Q.31. What is conglomerate?

Ans. A sedimentary rock composed of rounded fragments, stones or boulders cemented by lime and other cementing materials.

Q.32. What are the two types of inosilicates?

Ans. Single chain inosilicates - Eg. Pyroxene
Double chain inosilicates - Eg. Amphiboles

Q.33. Give examples for ferro magnesian minerals.

Ans. Olivine, Pyroxene, Amphiboles

Q.34. Give examples for Non-ferro magnesian minerals.

Ans. Quartz, Feldspar

Q.35. What are the fundamental soil forming processes?

Ans. Eluviation, Illuviation

Q.36. What are the specific soil forming processes?

Ans. Podzolization, Laterization, Humification, Calcification

Q.37. Who proposed soil formation as a function of climate, organisms, relief, parent material, time etc.

Ans. Jenny (1941)

Q.38. Write the equation of soil formation by Jenny?

Ans. S = f(cl, o, r, p, t....)

Q.39. What are active soil forming factors?

Ans. Climate and organisms

Q.40. What are the passive factors of soil formation ?

Ans. Parent material, relief and time

Q.41. Which soil forming factor has no relevance after soil reaches its maturity?

Ans. Time

Q.42. What are the soil forming processes prevalent in arid and semi- arid regions?

Ans. Calcification, gypsification, salinization and alkalization

Q.43. What is illuviation?

Ans. Accumulation of materials removed from the upper horizons at a depth beneath the soil surface.

Q.44. What is podzolization ?

Ans. Process of removal of oxides of iron, aluminium, humus and carbonates from the upper horizons.

Q.45. What is Eluviation ?

Ans. Mobilization and translocation of certain constituents like clay, silica, sesquioxides and humus etc. from one horizon to another.

Q.46. What is laterization ?

Ans. Process of removal of silica and accumulation of sesquioxides (Geothite, gibbsite etc) with or without the formation of iron stone and concretions.

Q.47. Which climatic condition is favourable for podzolization?

Ans. Cool and humid climatic conditions.

Q.48. Which parent materials are suitable for podzolization?

Ans. Silicious parent materials.

Q.49. Which climatic condition is favourable for laterization?

Ans. Warm and humid climatic conditions.

Q.50. What is Argillization?

Ans. Leaching of dispersed clay particles from the upper horizons and their accumulation in the lower horizons.

Q.51. What is Gleization?

Ans. Process of reduction, due to anaerobic condition of iron in waterlogged soils with the formation of mottles and concretions of iron and manganese.

Q.52. What is pedoturbation?

Ans. The process of intermixing of soil particles due to swelling and shrinking of clay particles.

Q.53. What is bisequem profile?

Ans. A soil profile having two sola one above the other.

Q.54. What is hydromorphic soil?

Ans. Soil developed under waterlogged condition.

Q.55. What are prairie soils?

Ans. A group of zonal soils formed in cool temperate to temperate humid climate under tall grass vegetation.

Q.56. What is silickensides?

Ans. Polished ped faces formed in soils having high amounts of smectite clay. Due to swelling and shrinking property, the gliding part of one ped over the other result in polished ped faces.

Q.57. Which soil forming process is responsible for the formation of gypsum from calcium sulphate?

Ans. Hydration.

Q.58. Which soil forming process is responsible for the formation of Allophane from orthoclase feldspar?

Ans. Hydrolysis

Q.59. What is soil horizon?

Ans. Layer of soil approximately parallel to the soil surface produced by the soil forming factors.

Q.60. What is the smallest volume that can be called as "a soil"?

Ans. Pedon.

Q.61. What is a pedon?

Ans. The smallest volume of soil that can be recognized as a soil individual.

Q.62. What is soil solum?

Ans. Soil horizons A and B, (A+B) are collectively called as soil solum.

Q.63. What is soil regolith?

Ans. Soil horizons A, B and C (A+B+C) are collectively called as soil regolith.

Q.64. What is Regolith?

Ans. The unconsolidated material of weathered rock and soil material overlying a solid rock on the earth's surface.

Q.65. What is the upper and biochemically weathered part of the regolith?

Ans. Soil

Q.66. What is polypedon?

Ans. A group of contiguous similar pedons that are bounded in all sides by not soil or by pedons of unlike characters.

Q.67. Who proposed the stages of soil development?

Ans. Mohr, Van Basen

Q.68. What are the different stages of soil development?

Ans. Initial, Juvenile, Virile, Senile, final are the different stages of soil development

Q.69. What is catena?

Ans. A sequence of soils developed from similar parent material under similar climatic conditions but varying relief and drainage conditions.

Q.70. What is toposequence ?

Ans. A sequence of related soils that differ from one another as a result of variations in topography.

Q.71. What is Lithosequence ?

Ans. A sequence of related soils that differ from one another in certain soil characteristics as a result of variations in parent material.

Q.72. What is Climosequence ?

Ans. A sequence of related soils that differ from one another in certain soil characterstics as a result of variations in climate.

Q.73. What is chronosequence?

Ans. A sequence of related soils that differ from one another in certain soil characterstics as a result of variations in time.

Q.74. What are ectodynamomorphic soils?

Ans. The soil with properties that have been influenced by soil forming factors other than parent material.

Q.75. What are endodynamomorphic soils?

Ans. The soil with properties that have been influenced primarily by the parent material.

Q.76. What is parent material?

Ans. The unconsolidated mass from which the solum develops by the soil forming factors or processes. The C horizon is referred to as the parent material of soils.

Q.77. What is lessivage?

Ans. Movement of clay particles in suspension.

Q.78. What is crotovinas?

Ans. An animal burrow filled with excrements or materials from other horizons.

Q.79. In which soil, bed rocks are obviously absent ?

Ans. Alluvial soils

Q.80. Which scale gives the hardness of minerals ?

Ans. Moh's scale of hardness measured by Scoloro meter. Talc - 1, Gypsum - 2, Calcite - 3, Fluorite - 4, Apatite - 5, Orthoclase feldspar - 6, Quartz - 7, Topaz - 8, Corundum - 9, Diamond - 10

Moh's scale of hardness of minerals is given as below

Talc	Mineral	Hardness
↓	↓	↓
Diamond		1 to 10

Q.81. What are the parameters that decide the stability of mineral to weathering and decomposition?

Ans. Degree of basicity, Degree of linkage to tetrahedron, Silica-Sesquioxide ratio.

Q.82. When bases are more, how would be the stability of minerals?

Ans. Increase.

Q.83. Which is the product of exhaustive hydrolysis and leaching?

Ans. Kaolinite

Q.84. Which is the dominant clay mineral in humid tropics?

Ans. Kaolinite

Q.85. Which are the clay minerals predominating in the soils of arid regions?

Ans. Illite and vermiculite.

Q.86. From which type of rocks, Kaolinite is derived from?

Ans. Sedimentary rocks.

Q.87. From which rock, montmorillonite is derived from?

Ans. Granite, Basalt, Gabbro.

Q.88. Name the unconsolidated materials from which the solum develops.

Ans. Parent material

Q.89. Name the materials transported and deposited by water.

Ans. Alluvium.

Q.90. Name the materials transported by water and settled out in lakes.

Ans. Lacustrine.

Q.91. Name the sand materials transported by wind.

Ans. Aeolian.

Q.92. What is Loess?

Ans. Silt particles transported by wind.

Q.93. What is soil catena?

Ans. A regular repetition of soil properties in association with certain relief, the parent materials may or may not be same, is called as soil catena.

Q.94. Name the soil formed under the influence of local conditions.

Ans. Intrazonal soils.

Q.95. Name the soils formed under the influence of climate.

Ans. Zonal soils.

Q.96. Which soil comes under Azonal group ?

Ans. Alluvial soils.

Q.97. Which soil order has slickensides and gilgai microrelief?

Ans. Vertisols.

Q.98. Which soil has honey-comb structure ?

Ans. Lateritic soils.

Q.99. What is the basis for early systems of soil classification?

Ans. Soil genesis.

Q.100. What is the basis for recent systems of soil classification?

Ans. Based on soil parentages and morphology.

Q.101. What are the categories of early systems of soil classification?

Ans. Zonal, Intrazonal and Azonal soils

Q.102. What is soil moisture regime?

Ans. Presence or absence of water held at a tension of < 1500 K Pa in the soil moisture control section during different periods of a year.

Q.103. Define soil series.

Ans. Soil series is a group of soils similar in their characteristics and arrangement of horizons.

Q.104. Who introduced the term soil catena?

Ans. Milne (1935).

Q.105. List out the early systems of soil classification.

Ans. Dokuchaiev genetic system (1900), Coffey's system of classification (1912), Marbut's morpho-genetic system (1935) and Baldwin genetic approach (1938).

Q.106. List out the merits of soil taxonomy.

Ans. Soil taxonomy is based on soil properties, unlike early systems of classification which is based on soil genesis. Soil taxonomy is an orderly scheme that facilitate to add new soil group, suborder and order.

Q.107. Write the categories of soil taxonomy.

Ans. Order, Suborder, Great group, Subgroup, Family, Series, Type and Phase.

Q.108. When was the 7th approximation published?

Ans. 1960 with supplements in 1964 and 1967.

Q.109. When was the soil taxonomy, a basic system of soil classification for making and interpreting soil surveys published?

Ans. 1975

Q.110. How many soil orders are there in the 7th approximation?

Ans. 12 (Twelve)

Q.111. What are the soil orders in 7th approximation?

Ans. Entisol, Vertisol, Inceptisol, Aridisol, Mollisol, Spodosol, Alfisol, Ultisol, Oxisol, Histosol, Andisol, Gelisol.

Q.112. Who proposed morphogenetic system of soil classification?

Ans. Marbut

Q.113. Who introduced the soil zonality concept?

Ans. Dokuchaiev

Q.114. Who proposed the weathering stages in soil formation?

Ans. Mohr

Q.115. Who proposed the weathering sequence?

Ans. Goldich (1938)

Q.116. Which is the common soil order found in warm humid regions?

Ans. Ultisols

Q.117. What is the soil order developed under grassland vegetation?

Ans. Mollisols

Q.118. What is the average organic matter content of peat soil?

Ans. More than 50 per cent

Q.119. Which soil order has cryoturbation or frost mixing?

Ans. Gelisol

Q.120. What is the base saturation percent of umbric horizon?

Ans. <50 Percent

Q.121. Which soil order is characterized by excess weathering with dominant iron and aluminium oxide clays and is common in tropics?

Ans. Oxisols

Q.122. What type of parent material is essential for podzolizalion?

Ans. Silicious parent material

Q.123. Which type of parent material is suitable for laterization?

Ans. Basic parent materials with sufficient iron bearing minerals.

Q.124. Whar are the main products of weathering?

Ans. Silica, alumina, soluble cations

Q.125. Which climatic condition hasten the rate of soil development?

Ans. Warm humid climate

Q.126. Name the prominent soil structure in black soil.

Ans. Angular blocky

Q.126. Name the typical soil for the temperate humid region.

Ans. Podzol

Q.127. Name the soils of semi arid and sub humid regions.

Ans. Chernozems

Q.128. Name the soil groups given by Dokuchaiev.

Ans. Pedocals and Pedalfers

Q.129. What is the fundamental unit in USDA soil classification?

Ans. Soil Profile

Q.130. In USDA soil classification, on what basis soil series are subdivided into soil types?

Ans. Soil Structure

Q.131. What is the soil classification followed in India?

Ans. USDA soil classification

Q.132. What is the depth of diagnostic horizons or epipedons from the soil surface?

Ans. 15 - 18 cm

Q.133. What would be the gypsum content in gypsic horizon?

Ans. 5 per cent

Q.134. What is the minimum per cent of salt in salic subsurface diagnostic horizon?

Ans. 2 per cent

Q.135. In defining soil temperature regimes, at what depth the temperature is mentioned?

Ans. 50 cm of soil depth, lithic contact, paralithic contact

Q.136. In which soil, gilgai micro relief is a diagnostic feature?

Ans. Black soils

Q.137. Which soil order has argillicpedoturbation?

Ans. Vertisols

Q.138. Which soil order includes soils developed from volcanic eruption?

Ans. Andosols

Q.139. Which is the diagnostic surface horizon with high SAR?

Ans. Natric.

Q.140. What is plinthite?

Ans. A non indurated mixture of iron and aluminium oxides, clay, quartz and other diluents that commonly occurs as red soil mottles usually arranged in platy, polygonal or reticulate patterns.

Q.141. What is the primary source of salts in salt affected soils?

Ans. Weathering of rocks

Q.142. What is the indurated surface horizon cemented by carbonates?

Ans. Petrocalcic horizon.

Q.143. What is the indurated surface horizon cemented by Gypsum?

Ans. Petrogypsic horizon.

Q.144. Give an example for non-silicate secondary mineral.

Ans. Gibbsite

Q.145. Name some secondary minerals observed in soils of arid and semi-arid regions.

Ans. Gypsum, Calcite, Apatite

Q.146. Name the rocks with high amount of secondary mineral.

Ans. Sedimentary rocks

Q.147. Name the most common clay mineral observed in soils.

Ans. Montmorillonite.

Q.148. What is the indication of quick chemical change in rocks?

Ans. Presence of Haematite in rocks

Q.149. Name the mineral that is abundantly found in highly weathered soils of the tropical region.

Ans. Gibbsite

Q.150. Name an important mineral in water logged soils.

Ans. Siderite

Q.151. What is the quantity of essential minerals in the rocks?

Ans. 95-98%

Q.152. What is the quantity of accessory minerals in the rocks?

Ans. 2-5%

Q.153. What is the specific gravity of heavy minerals?

Ans. More than 2.85 g cm^{-3}

Q.154. Give examples for heavy minerals

Ans. Olivine, Haematite, Pyrite and Augite

Q.155. What is the specific gravity of light minerals?

Ans. Less than 2.85 g cm^{-3} Eg. quartz, feldspar, muscovite

Q.156. Give an example for hard and compact rock.

Ans. Basalt

Q.157. Give an example for loose and feebly aggregated rock.

Ans. Conglomerate and Sandstone

Q.158. What is the silica content in acidic rocks?

Ans. More than 60%

Q.159. What is the silica content in basic rocks?

Ans. Less than 50%

Q.160. What is the silica content in neutral rocks?

Ans. 51-60%

Q.161. By which metamorphism process, granites are converted into gneiss?

Ans. Thermal metamorphism

Q.162. By which metamorphism process, gneisses are converted into schists?

Ans. Dynamo-thermal metamorphism

Q.163. Name some minerals formed during the advanced stage of weathering.

Ans. Kaolinite and Halloysite

Q.164. Who proposed the stability index of minerals?

Ans. Jackson

Q.165. What is the K_2O content in lllite clay mineral?

Ans. 6 - 10%

Q.166. What imparts very dark colour in black soil?

Ans. Clay-humus complex or presence of titanium, Magnetite

Q.167. Name some minerals formed at the initial stage of weathering?

Ans. Gypsum, Calcite, Olivine, Hornblende, Pyroxene, biotite and glauconite

Q.168. Mention the physical weathering agents.

Ans. Temperature, Water, Wind

Q.169. Define Physical weathering.

Ans. Physical weathering is the disintegration of rocks into smaller particles without any chemical change or formation of new products.

Q.170. Define Chemical weathering.

Ans. It is the transformation of original rocks into new compounds having different physical properties and chemical composition.

Q.171. Mention the chemical weathering processes.

Ans. Hydration, Hydrolysis, Solution, Carbonation, Oxidation and Reduction.

Q.172. Define biological weathering.

Ans. It is the disintegration of rocks into smaller particles by biological agents such as man, animals, micro organisms, roots of higher plants.

Q.173. How are the orders differentiated?

Ans. Based on the presence or absence of diagnostic horizons.

Q.174. Which soil order consists of young soils with little horizonation?

Ans. Entisols

Q.175. Which soil order has ochric epipedon?

Ans. Aridisols

Q.176. What are the characteristic diagnostic horizons in Aridisols?

Ans. Calcic, Gypsic or Salic

Q.177. Which soil order has argillic subsurface horizon?

Ans. Alfisols

Q.178. Which soil order has low (< 35%) base saturation?

Ans. Ultisols

Q.179. Which soil order has highly weathered soils?

Ans. Oxisols

Q.180. Which soil order consists of Organic soils viz., bog, peat and muck?

Ans. Histosols

Q.181. What is the origin of soils in Andisols?

Ans. Volcanic eruption

Q.182. What is the characteristic mineral in Andisols?

Ans. Amorphous minerals like Allophane, Imogolite.

Q.183. What is the characteric feature of Gelisols?

Ans. Cryoturbation or frost mixing

Q.184. What are the Epipedons or diagnostic surface horizons?

Ans. Mollic, Umbric, Anthropic, Histic, Plaggen, Ochric.

Q.185. Mention the diagnostic subsurface horizons or endopedons

Ans. Agric, Albic, Argillic, Calcic, Natric, Spodic, Oxic, Sombric, Cambic, Kandic, Sulphuric, Salic, Placic, Glossic, Gypsic, Duripan, Fragipan, Petro calcic, Petrogypsic.

Q.186. How many epipedons and endopedons are there?

Ans. 9 epipedons and 19 endopedons.

Q.187. What are the hardpan horizons in subsurface?

Ans. Duripan, Fragipan, Plinthite, Petrocalcic, Petrogypsic, placic horizons.

❑❑❑

Chapter-2

SOIL SURVEY AND INTERPRETATION

Q.1. Expand NBSS and LUP. Where is its head quarters?

Ans. National Bureau of Soil Survey and Land Use Planning. Nagpur

Q.2. What were the maps prepared by NBSS and LUP after national survey project between 1986 and 1996?

Ans. Soil resource map of India on 1:1 Million scale and soil resource map of the states and union territories on 1:2,50,000 scale.

Q.3. Define soil survey.

Ans. Soil survey is the systematic examination, description, classification and mapping of soils of an area in their natural environment.

Q.4. What is the fundamental purpose of soil survey?

Ans. Understanding the soils of an area in terms of their properties, genesis, classification and nomenclature.

Q.5. What is the applied purpose of soil survey?

Ans. Transfer of technology, developing optimum land use plans, delineating degraded soils and suggest conservation measures, tax appraisal, designing infrastructure etc.

Q.6. What is the type of survey used for mapping small areas such as watershed or a research station?

Ans. Grid survey

Q.7. Which method of survey is recommended when there are changes in physiography and other surface feature?

Ans. Free survey

Q.8. Define soil delineation.

Ans. An area completely circumscribed by a soil boundary on a map.

Q.9. What is the aggregation of soil delineations identified by a unique symbol on a map called as?

Ans. Soil map unit

Q.10. Mention the types of soil survey.

Ans. Reconnaissance, Detailed, Detailed - Reconnaissance, Semi detailed, Exploratory, Rapid Reconnaissance.

Q.11. Which soil survey is used to prepare resource inventory of large area?

Ans. Reconnaissance soil survey

Q.12. What is the frequency of field observations in rapid reconnaissance soil survey?

Ans. 1 / 10,000 ha.

Q.13. What is the frequency of field observation in low intensity detailed soil survey?

Ans. 1 / ha

Q.14. What is the frequency of field observation in high intensity detailed soil survey?

Ans. 4 / ha

Q.15. In detailed soil survey, at what interval the soils are examined?

Ans. 0.25 - 0.5 km

Q.16. What are the base materials used in the preparation of soil maps for detailed soil survey?

Ans. Cadastral maps, Aerial photographs, Satellite imagery

Q.17. What is the scale of cadastral map used in detailed soil survey?

Ans. 1: 8,000, 1: 4,000

Q.18. At what interval, pedons are examined in reconnaissance soil survey?

Ans. 3.0 - 6.0 km

Q.19. What is the scale of aerial photograph used in detailed soil survey?

Ans. 1:15,000

Q.20. What are benchmark soils?

Ans. Soils occupying a key interpretative position in soil classification framework and or covering large area is called benchmark soils.

Q.21. Mention the base maps used in soil survey.

Ans. Cadastral maps (1: 4,000 - 1: 8,000 scale), Toposheets (1: 25,000-1:2,50,000)

Q.22. Expand GSI. Where is its headquarters?

Ans. Geological Survey of India, Kolkata.

Q.23. What is the scale of base maps used in RSS?

Ans. 1: 1,00,000 to 1: 2,50,000

Q.24. What is the base map used in RSS?

Ans. Toposheets, aerial photographs, satellite imagery

Q.25. What is the scale of base map in rapid reconnaissance Soil survey?

Ans. 1:1,000,000 or smaller

Q.26. What is the mapping unit in RSS?

Ans. Phases of association of soil families

Q.27. What is the purpose of semi detailed survey?

Ans. For district level planning

Q.28. What is the scale of base map in semi detailed survey?

Ans. 1: 5,000

Q.29. What is the scale of base map in detailed soil survey?

Ans. 1: 4,000 in High intensity survey
1: 8,000 in Low intensity survey

Q.30. What is the mapping unit in Detailed soil survey?

Ans. Phases of soil series

Q.31. What is the purpose of Detailed soil survey?

Ans. For village or watershed planning

Q.32. Define Land suitability evaluation.

Ans. It is the process of assessing the suitability of land for specific use.

Q.33. What is Land Capability Classification?

Ans. Land capability classification is the grouping of land units into defined classes based on their capability.

Q.34. What are the different capability subclasses?

Ans. Capability subclasses are based on kinds of dominant limitation viz., wetness(w), climate(c) soil (s) and erosion(e).

Q.35. What does a land capability unit denote?

Ans. It is a subdivision of land capability subclass which includes soils which are uniform in their characteristics, potential and limitations and require uniform conservation treatment and management practices.

Q.36. Who proposed the criteria for Land Capability Classification?

Ans. Klingebiel and Montgomery (1961)

Q.37. What are the criteria used for assessing land suitablity evaluation?

Ans. Inherent soil characteristics, external land features. Environmental factors that limit land use.

Q.38. What are the three categories in LCC?

Ans. Capability classes, capability subclasses and capability units.

Q.39. How many capability classes are present?

Ans. Eight

Q.40. Which capability class soils are highly suited for crop production?

Ans. Class I

Q.41. Mention the standard colours used to denote different capability classes.

Ans. Class I - Green Class V - Dark grey
Class II - Yellow Class VI - Orange
Class III - Brown Class VII - Red
Class IV - Pink Class VIII - Purple

Q.42. Which are the capability classes suitable for cultivation?

Ans. Class I to IV

Q.43. Which are the capability classes suitable for grazing and forestry?

Ans. Class V to VII

Q.44. What is the utility of capability class VIII in LCC?

Ans. Non arable and suitable for wild life or recreation.

Q.45. Which capability class has no subclass?

Ans. Class I since it has no limitations.

Q.46. What is land irrigability classification?

Ans. It is evaluating the suitability of soils for sustained use under irrigation.

Q.47. What are the criteria used for evaluating soil suitability for irrigation by FAO?

Ans. Soil texture, depth, permeability, salinity, topography.

Q.48. How many soil irrigability classes are there in FAO classification?

Ans. Six

Q.49. Who proposed the capability index rating approach for evaluating soil irrigability?

Ans. Sys (1976)

Q.50. How are the soils rated based on capability index

Ans.

Capability	Index rating	Notation for suitability
>80	S1	Very suitable
60-80	S2	Moderately suitable
40-60	S3	Marginally suitable
20-40	N1	Almost unsuitable
<20	N2	Unsuitable

Q.51. Who proposed the different systems of land evaluation?

Ans. Storie (1954), Riquier et al. (1970) FAO (1976), Sys (1976)

Q.52. What are the land evaluation classification categories in FAO system (1976)?

Ans. Orders, Classes, Subclasses, Units

Q.53. How many orders and classes are there in FAO system of land evaluation?

Ans. Two orders S - Suitable N - Not suitable
Three classes under order S - (S1, S2, S3)
Two classes under order N - (N1, N2)

Q.54. What are the subclasses in FAO system of land evaluation?

Ans. Based on the kind of limitations.

c : Climate
t : Topography
w : Wetness
n : Salinity / sodicity
f : Soil fertility
s : Soil physical conditions (influencing soil water relationship and management)

Q.55. What are the steps in soil site suitability evaluation?

Ans.

1. Collecting the soil site characteristics that influence the productive capacity of the particular land utilisation type (LUT) under consideration.
2. Determination of soil and climatic requirements of the considered LUT.
3. Comparing the collected characteristics with the standard requirements of the considered LUT.

Q.56. Who is the author of the book "A Text Book of Pedology - Concept and Application" ?

Ans. J. Sehgal (1996), Revised Edition (2005)

❏❏❏

Chapter-3

SOIL PHYSICS

Q.1. Who is the father of soil physics ?

Ans. Daniel Hillel

Q.2. What is the number of classification in size groups of soil particles by International Society of Soil Science (ISSS)?

Ans. Four

Q.3. What are soil separates?

Ans. Stones, gravels, sand, silt, clay

Q.4. Which are the primary soil particles?

Ans. Clay, silt, sand

Q.5. What is the size of silt particles in 1SSS classification of soil texture?

Ans. 2-20 μ

Q.6. Name the method of textural analysis of all soils except calcareous and organic soils.

Ans. Bouyoucos Hydrometer method

Q.7. Name the method of accurate analysis of soil texture.

Ans. International pipette method based on Stoke's law

$$V = \frac{2}{9}\frac{(dp - d)\, gr^2}{\eta}$$

Q.8. Which soil has high plasticity and stickiness?

Ans. Clay soils

Q.9. In which soils, water infiltration is more rapid?

Ans. Sandy soils

Q.10. What is soil texture?

Ans. The relative proportion of soil particles is called as soil texture.

Q.11. What is sticky limit?

Ans. Water content above that a mixture of soil and water will adhere to a steel spatula

Q.12. What is plasticity index?

Ans. Difference between the lower and upper plasticity limits of soil consistency is called as plasticity index

Q.13. Name the element essential for the flocculation of soil particles.

Ans. Calcium

Q.14. Name the element that cause deflocculation of soil particles.

Ans. Sodium

Q.15. Name the element used to reduce Zeta potential of soil particles.

Ans. Calcium

Q.16. What will occur if zeta potential of soil particles increases?

Ans. Deflocculation

Q.17. Name the substances that cause cementation of soil particles.

Ans. Cations, oxides of iron and aluminium, anions, humic substances.

Q.18. What are the direct methods of soil structure characterization and evaluation ?

Ans. Microscopic methods

Q.19. What are the indirect methods of soil structure characterization and evaluation?

Ans. Wet and dry sieving methods, mean weight diameter, stability index and structural co-efficient.

Q.20. What is soil structure ?

Ans. The arrangement of primary particles and their aggregates into a certain definite pattern

Q.21. What are the types of soil structure?

Ans. Plate like, Prism like, Block like, Spheroidal.

Q.22. For better soil structure, how should be the value of cumulative mean weight diameter and structural coefficient?

Ans. Cumulative mean weight diameter should be lower and structrual coefficient should be higher

Q.23. What are the structural sub types of soil?

Ans. Plate like - Platy, laminar

Prism like - Prismatic, columnar

Block like - Angular blocky, Sub angular blocky

Spheroidal - Sphere like - Granular, crumb

Q.24. What is soil tilth?

Ans. The physical condition of soil in relation to plant growth

Q.25. Which is an optimum soil condition for the tillage operation to produce aggregate of suitable sizes?

Ans. Friable

Q.26. What are the grades of soil structrue?

Ans. Strucureless, Weak, Moderate, Strong

Q.27. What is infiltration rate?

Ans. The distance travelled by water through soil column per unit time.

Q.28. What is the factor that influences the infiltration of water?

Ans. Porosity

Q.29. What are the factors that influence infiltration rate?

Ans. Soil structure, porosity, initial soil water content

Q.30. What is the unit of atmospheric pressure ?

Ans. Dynes cm^{-2}

Q.31. What is hydraulic conductivity?

Ans. Rate of flow of liquid through a porous medium under unit hydraulic gradient

Q.32. In the laboratory, how is hydraulic conductivity measured?

Ans. Using constant head permeameter

Q.33. In the field, how is hydraulic conductivity measured?

Ans. Piezometer, auger hole method

Q.34. What is the unit of surface tension?

Ans. Dynes cm^{-1}

Q.35. What is matric potential?

Ans. The potential resulting from adsorptive and capillary forces of the soil matrix acting on soil water is called as matrix potential.

Q.36. What is gravitational potential?

Ans. The potential resulting from the earth's gravitational force is called as Gravitational potential.

Q.37. What is solute or osmotic potential?

Ans. The potential resulting from the presence of solutes in soil water is called as osmotic potential.

Q.38. What is the soil water potential at saturation in soil?

Ans. 0 bars

Q.39. What is the soil water potential at air dryness?

Ans. -15 bars

Q.40. What is the soil water potential at field capacity?

Ans. 33 Pa

Q.41. What is the soil water potential at permanent wilting point?

Ans. 1500 Pa

Q.42. What is the soil water potential for plant available water?

Ans. 10 - 33 Pa, 33 - 1500 Pa

Q.43. Which instrument is used for the measurement of negative soil water pressure (suction)?

Ans. Tensiometer

Q.44. Upto which depth, Tensiometer can be used to measure soil water pressure?

Ans. 800 - 900 cm

Q.45. Which apparatus is used to measure soil water potential in the field?

Ans. Tensiometer

Q.46. In the laboratory, which apparatus is used to measure soil water potential?

Ans. Porous cup apparatus, pressure plate apparatus

Q.47. What is aeration porosity?

Ans. Percentage of pores which is drained by 100cm of suction

Q.48. Which blocks are used in electrical resistance method of soil water estimation?

Ans. Gypsum blocks

Q.49. What is the source of gamma radiation in gamma-ray scanner used for soil water estimation?

Ans. $^{137}C_s$

Q.50. What is the percentage of N_2 in atmospheric air?

Ans. 79. 0 %

Q.51. What is the percentage of carbon dioxide in soil air?

Ans. 0.3 %

Q.52. By which process, more than 90 per cent of gaseous exchange in soil takes place?

Ans. Diffusion

Q.53. By which process, diffusion of gases under total gas pressure gradient takes place?

Ans. Mass flow

Q.54. By which process, the heat absorbed by soil during day on sunny days is lost during night?

Ans. Radiation

Q.55. What does the dominant spectral colour in Munsell colour chart indicate?

Ans. Hue

Q.56. In soil colour determination, what is hue related to?

Ans. Wavelength of light or dominant spectral colour

Q.57. What does relative purity or strength of spectral colour in Munsell colour chart indicate?

Ans. Chroma

Q.58. In soil colour determination, what is chroma related to?

Ans. Purity of light

Q.59. In soil colour determination, what is value related to?

Ans. Intensity of light

Q.60. What are the five principle hues in Munsell colour chart?

Ans. Red, Yellow, Green, Blue and Purple.

Q.61. What are complimentary colours according to colour theory?

Ans. Two colors are complimentary when mixed in the proper proportion, they produces natural color (White + Black = Grey)

Q.62. What is range of value in Munsell colour chart?

Ans. 1-10

Q.63. What are the primary colors?

Ans. Red, Green, Blue.

Q.64. In Munsell colour chart, what does hue number 9 denote?

Ans. Absolute yellow

Q.65. In Munsell colour chart what does value number one denote?

Ans. Absolute black

Q.66. In Munsell colour chart, what does value number 10 denote?

Ans. Absolute white

Q.67. What is the range of chroma in Munsell colour chart?

Ans. 0 - 8

Q.68. By which process diffusion of gases under the concentration gradient takes place?

Ans. Root interception

Q.69. How can Swelling and shrinking properties of soil be determined by?

Ans. Co-efficient of Linear Extensibility (COLE), Potential Volume Change (PVC), Swell index.

Q.70. In soil, which instrument is used to measure water pressure head or hydraulic head under saturated condition?

Ans. Piezometer

Q.71. In soil, which instrument is used to measure water pressure head or hydraulic head at unsaturated condition?

Ans. Tensiometer

Q.72. What is soil consistency?

Ans. The degree of plasticity and stickiness of a soil

Q.73. What will be the effect of incorporating crop residues into the soil?

Ans. Improves soil structure

Q.74. Give Torricelli's equation.

Ans. $V = 2g^2$

Q.75. What is the unit of oxygen diffusion rate (ODR)?

Ans. g / cm^2 / min

Q.76. What is the bulk density of organic matter?

Ans. 0.5 g / cc

Q.77. What should be the air content in soil for favourable plant growth?

Ans. 10 -12%

Q.78. What is the upper limit of available water in soil?

Ans. Field capacity

Q.79. What is the unit of thermal diffusivity?

Ans. cm / sec^2

Q.80. What is the critical limit of oxygen diffusion rate?

Ans. 20×10^{-8}g / cm^2 min

Q.81. What is the unit of specific heat?

Ans. Calories / g

Q.82. What is the specific heat of soils?

Ans. 0.2 Cal / g

Q.83. What is the specific heat of minerals?

Ans. 0.11 Cal / g

Q.84. What is the specific heat of Organic Matter?

Ans. 0.44 Cal / g

Q.85. What is the Unit of thermal conductivity?

Ans. °C / cm / sec

Q.86. What will happen to the thermal conductivity when compactness and wetness of soil increase?

Ans. Increased

Q.87. What is the unit of viscosity?

Ans. Poise

Q.88. Who proposed pF scale?

Ans. R.K. Schofield

Q.89. What is the value of pF scale at field capacity?

Ans. 2.53

Q.90. What are the ill effects of soil compaction?

Ans. Increase in bulk density, decrease in hydraulic conductivity.

Q.91. Define soil porosity

Ans. The ratio of total volume of pore space to the total volume of soil is called porosity.

Q.92. What is the index of the relative volume of soil pores?

Ans. Void ratio

Q.93. Define void ratio.

Ans. The ratio of total volume of pores to the total volume of soil solids is called as Void ratio.

Q.94. Define air filled polority.

Ans. The ratio of volume of air space to the total volume of soil is called as air filled porosity. It is the measure of the relative air content of the soil.

Q.95. How will you evaluate air filled porosity?

Ans. By air space pycnometer

Q.96. How will you determine air permeability of soil under field and laboratory conditions?

Ans. By using air permeameter and garometer.

Q.97. What is bulk density and particle density?

Ans. Bulk density is defined as the weight per unit volume of soil including pore space. Particle density is defined as the weight per unit volume of soil solids alone excluding pore space. The unit of density is gcc^{-1} or Mgm^{-3}.

Q.98. How will you determine soil bulk density?

Ans. By core sampler method and excavation method.

Q.99. What should be the air capacity of soil?

Ans. Air capacity of the soil should be greater than 10 % for all crops except paddy.

Q.100. How will you determine the bulk density of entire profile?

Ans. By radiation $^{60}C_o$.

□□□

Chapter-4

SOIL CHEMISTRY AND CLAY MINERALOGY

Q.1. What is the percentage of organic carbon in organic matter?

Ans. 58%

Q.2. What is the average C:N ratio of Indian soils?

Ans. 10 - 12:1

Q.3. What is the SI unit of cation exchange capacity?

Ans. $Cmol(P+)Kg^{-1}$

Q.4. Who proposed the term humic acid ?

Ans. Berzelius

Q.5. What is the equation, pH-l / 2pCa = Constant called as?

Ans. Lime potential

Q.6. On the basis of Bronsted - Lowry concept, what does an acid denote?

Ans. Protons

Q.7. Osmotic pressure = ECx ?

Ans. 0.36

Q.8. What is soil pH ?

Ans. Measure of Hydrogen ion concentration in soil solution.

Q.9. What are the two types of soil acidity?

Ans. Active acidity and Exchange or Reserve or Replaceable acidity.

Q.10. What is known as quick lime?

Ans. CaO

Q.11. Write the units of electrical conductivity (EC).

Ans. Micro mhos / cm, Milli mhos / cm, dS / m

Q.12. For a good water sample, what should be the SAR value?

Ans. <10

Q.13. Who is the author of the book 'Principles of Soil Chemistry?

Ans. Kim H. Tan

Q.14. What is the predominant form(s) of phosphorus and sulphur in soil?

Ans. Organic

Q.15. Name the primary inorganic form of phosphorus in soil.

Ans. Apatite

Q.16. What are the two main sources of inorganic sulphur in soil?

Ans. Gypsum and pyrites

Q.17. What is the method used to estimate total carbon in soil?

Ans. Dry combustion method

Q.18. Organic matter in surface soil = organic carbon x ________?

Ans. 1.724

Q.19. Organic matter in sub surface soil = organic carbon x ___?

Ans. 2.54

Q.20. Which climatic conditions are conducive for the accumulation of organic matter in soil?

Ans. Low temperature and high rainfall

Q.21. Name the insoluble fraction of organic matter.

Ans. Humin

Q.22. What is the upper limit of colloidal particles?

Ans. 2 microns

Q.23. What is the cation exchange capacity of humus?

Ans. 250 - 1500 $Cmol(P+)Kg^{-1}$

Q.24. Who discovered ion exchange phenomenon?

Ans. Thompson and Way

Q.25. Name the resin used for softening of water.

Ans. Sodium saturated exchange resins

Q.26. What are the elements responsible for the hardness of water?

Ans. Ca and Mg

Q.27. Write freundlich's equation.

Ans. $X / m = KCn$

Q.28. Write Langmuir equation.

Ans. $X / m = K_1K_2C / 1 + K_2C$

Q.29. What is the equation used for heterovalent exchange reactions?

Ans. Gapon's equation

Q.30. Write Gapon's equation

Ans. K = NB / NA xmA / (mB) 1 / 2

Q.31. Write the formula to calculate Base Saturation Percentage

Ans. $$BSP = \frac{\text{Exchangeable cations}}{\text{CEC}} \times 100$$

where CEC is the cation exchange capacity

Q.32. Name the cation in soil complex that cause dispersion or focculation of soil particles.

Ans. Sodium

Q.33. Name the cation in soil complex that cause aggregation of soil particles.

Ans. Calcium

Q.34. Name the amendment used for the reclamation of acidic soils.

Ans. Lime ($CaCO_3$)

Q.35. Who proposed the term pH?

Ans. Sorensen

Q.36. What is the range of pH scale?

Ans. 1-14

Q.37. What is used to measure the EMF of a cell ?

Ans. Reference electrode

Q.38. At pH 5.0 and above, what is the main source (s) of hydrogen ions in soil?

Ans. Exchangeable H+

Q.39. In which soils buffering capacity is high?

Ans. Soils with 2:1 clay minerals

Q.40. What would be the pH value if the soil is alkaline?

Ans. > 7.0

Q.41. Which amendment is used for the reclamation of sodic or alkali soils?

Ans. Gypsum

Q.42. What is the ionic form of phosphorus when the soil pH is more than 9.0?

Ans. PO_4^{3-}

Q.43. At the pH range of 2.0 - 5.0, which ions in soil fix phosphates?

Ans. Iron and aluminium

Q.44. Which soils have high Phosphate fixing capacity?

Ans. Black soils

Q.45. Give the Avogadro's number.

Ans. 6.022×10^{23}

Q.46. What is the ionic radius of Al^{3+} ion?

Ans. 0.5° A

Q.47. What is Normality ?

Ans. Number of equivalents of a substance dissolved in 1 litre of a solution.

Q.48. Give Nernst equation

Ans. $(E = RT / nFlnK / M_n)$

Q.49. What is the diameter of the water molecules?

Ans. 3°A

Q.50. What is energy released during hydration of clay particles known as?

Ans. Heat of wetting

Q.51. Who introduced the term water potential?

Ans. Buckingham (1907)

Q.52. Who introduced the term capillary potential?

Ans. Hillel (1972)

Q.53. Under normal field conditions, what will be the soil water potential?

Ans. Negative

Q.54. Who proposed the law of mass action?

Ans. Guldberg

Q.55. What is the dissociation constant of water?

Ans. 1.8×10^{-14}

Q.56. Name the equation to calculate Ion activity coefficient.

Ans. Debye - huckel equation

Q.57. Who proposed the concept of ionic strength?

Ans. Lewis and Randall

Q.58. What is the lower size limit of colloid ?

Ans. 50° A

Q.59. What is the conversion factor for the conversion of nitrogen into protein?

Ans. Protein = N x 6.25

Q.60. What is the humus content of humic acid?

Ans. 3.5 %

Q.61. What is the method useful in identifying the different types of humic fractions ?

Ans. Infrared spectroscopy

Q.62. According to flaig hypothesis, what is the source of formation of fulvic and humic acids?

Ans. Lignin

Q.63. What is the common formula for phyllosilicates?

Ans. Si_2O_5

Q.64. What is the common formula for cyclosilicates?

Ans. SiO_3, Si_2O_5

Q.65. What is the common formula for inosilicates?

Ans. SiO_3, SiO_{11}

Q.66. What is the common formula for tectosilicates?

Ans. SiO_2

Q.67. Give examples for neosilicates.

Ans. Zircon, topaz

Q.68. Give examples for chain structured soil clay minerals.

Ans. Polygorskite, Sepiolite

Q.69. Give an example for 2 : 2 type of clay minerals.

Ans. Chlorite

Q.70. What is the name of subgroup that has Mg^{2+} ions in all the octahedral position?

Ans. Trioctahedral subgroup

Q.71. What is the name of subgroup that has Al^{3+} ions in all the octahedral position?

Ans. Dioctahedral subgroup

Q.72. Give an example for expanding 1:1 clay mineral.

Ans. Halloysite

Q.73. Which 1:1 clay mineral is the most widely distributed one in Indian soils?

Ans. Kaolinite

Q.74. Which soil orders have Kaolinite as important fraction of clay?

Ans. Alfisols and Vertisols

Q.75. Write the weathering sequence in the formation of soil clay minerals.

Ans. Igneous rock -Montmorillonite -Halloysite -Metahalloysite - Kaolinite

Q.76. What is the name of commercial grade montmorillonite?

Ans. Bentonite

Q.77. Which montmorillonite group of clay minerals, contain high Al content?

Ans. Beidellite

Q.78. What is the basal spacing of oven-dried (105°C) montmorillonite?

Ans. 10° A

Q.79. What is the cation exchange capacity of illite?

Ans. 10 meq / 100g

Q.80. In soils affected by low precipition, what would be the alteration of the illite mineral?

Ans. Illite altered to Montmorillonite / Kaolinite

Q.81. Which clay mineral has more tetrahedral layer isomorphic substitution?

Ans. Vermiculite

Q.82. Among the inorganic colloids, which has the largest cation exchange capacity?

Ans. Vermiculite

Q.83. In which clay mineral, potassium and ammonium fixation is more?

Ans. Vermiculite

Q.84. What is the value of X-ray diffraction peak for vermiculite?

Ans. 14.0°A

Q.85. In which soil orders, Vermiculite usually occurs as secondary minerals in the clay fractions?

Ans. Ultisols, Mollisols and Aridisols

Q.86. Which is considered as regularly interstratified chlorite montmorillonite clay?

Ans. Corrensite

Q.87. What is the basal spacing of Corrensite?

Ans. 26.4 - 31.0° A

Q.88. What are the types of minerals distinguished in the category of silica minerals?

Ans. Quartz,Tridymite and Crystobalite

Q.89. Give an example for hexagonal silica mineral.

Ans. Tridymite

Q.90. Give an example for trigonal silica mineral.

Ans. Quartz

Q.91. What is the surface area of silica minerals?

Ans. 2 - 3 m^2 / g

Q.92. At which soil pH, silica will dissolve?

Ans. > 9.0

Q.93. Write the sequential order of solubility of silica.

Ans. Mineral Quartz < Opal <Crystobalite <Amorphous silica

Q.94. Which is called as biogenic silica since it is mostly originated from plant?

Ans. Opal

Q.95. What is the basal spacing of quartz?

Ans. 3.34^{o} A

Q.96. For the calibration of DTA instrument, which is often used as a stable reference material ?

Ans. Quartz

Q.97. In which soil order Quartz is usually an important mineral?

Ans. Spodosols

Q.98. In which soil order, Quartz may be absent?

Ans. Oxisols

Q.99. Among the silica minerals, which one is considered as volcanic origin?

Ans. Crystobalite

Q.100. Give an example of anhydrous ferric oxide mineral.

Ans. Maghemite

Q.101. Give an example of a hydrated ferric oxide mineral.

Ans. Ferrihydrite

Q.102. Which is the most common aluminium hydrous oxide mineral in soil?

Ans. Diaspore

Q.103. Which aluminium hydrous oxide mineral is also called as hydragillite?

Ans. Gibbsite

Q.104. In gibbsite, which bond holds the two crystal layers together?

Ans. H-bonds

Q.105. What is the adsorption related to a covalent type of bonding of ions?

Ans. Specific adsorption of phosphate ions and heavy metals

Q.106. What is the X ray diffraction value peak for gibbsite and geothite?

Ans. Gibbsite: 4.82°A : Geothite : 4.18°A

Q.107. Which soil mineral imparts reddish-brown colour to the soils?

Ans. Haematite

Q.108. Which is the major mineral in highly weathered Ultisols and Oxisols of the humid tropical and subtropical regions?

Ans. Gibbsite

Q.109. Give an example of iron oxide mineral in soils.

Ans. Geothite

Q.110. What is the molecular ratio of Al: Si of allophane?

Ans. 1:1 or 1:1.5

Q.111. What is the chemical formula of allophane?

Ans. $SiO_2.Al_2\ O_3.2\ H_2O$

Q.112. What is the structure of imogolite?

Ans. Para crystalline

Q.113. What is the CEC of allophane?

Ans. 20 - 30 C mol (P+) Kg^{-1}

Q.114. What is the CEC of Imogolite?

Ans. 135 C mol (P+) Kg^{-1}

Q.115. What is the Anion exchange capacity of allophane?

Ans. 196 C mol (P+) Kg^{-1}

Q.116. Name the interaction processes of minerals with organic compounds.

Ans. Complex formation or chelation

Q.117. Name the method commonly used in the identification of amorphous mineral.

Ans. Differential Thermal Analysis (DTA)

Q.118. Name the standard reference material used in the Differential Thermal Analysis (DTA) method of clay mineral identification.

Ans. Calcined Kaolinite

Q.119. Name the method of qualitative analysis of clay mineral identification.

Ans. X-ray diffraction analysis

Q.120. Write a merit of X - ray diffraction analysis.

Ans. Non - destructive method

Q.121. What are the methods used to distinguish between expanding and non-expanding minerals in X - ray diffraction analysis?

Ans. K - saturation and Mg - saturation

Q.122. What procedure is followed to distinguish between Vermiculite, Chlorite and Kaolinite in X - ray diffraction analysis?

Ans. Heating at 500°C

Q.123. Which is called as siloxane bond?

Ans. Si-O-Si

Q.124. In which clay mineral, Siloxane bond (Si-O-Si) is present?

Ans. 2:1 clay mineral

Q.125. Which chemical compounds are commonly used for the measurement of surface areas of clay?

Ans. Water, N_2 gas and Acetyl pyridinum

Q.126. Which ions replace Al^{3+} ions in the crystal layer of clay?

Ans. Mg^{2+}, Fe^{3+} and Si^{4+}

Q.127. What is the type of charge for 1:1 clays?

Ans. pH - dependent charge

Q.128. Which ions are called as potential determining ions?

Ans. H+ and OH-

Q.129. What is the formula used for the calculation of negative charges?

Ans. Negative charges pH = pH H_2O - pH KCl

Q.130. What is surface potential?

Ans. Presence of opposite charges on the colloidal surface and in liquid phase develops an electrical potential at the solid - liquid interface.

Q.131. Define Helmholtz double layer theory.

Ans. Electrochemical potential at the colloidal surface drops linearly with increasing distance from the surface with in the double layer.

Q.132. If the valency of the ion increases, what happens to the thickness of double layer?

Ans. Decreases

Q.133. If the concentration of the bulk solution decreases, what will happen to the thickness of double layer?

Ans. Increases

Q.134. Among the adsorption equation, which equation describes adsorption processes in relation to surface tension?

Ans. Gibbs equation

Q.135. What is the type of reaction that occurs when the ligand donates electron pairs to a metal ion?

Ans. Coordination reaction

Q.136. Which clay mineral has high heat of adsorption?

Ans. Kaolinite

Q.137. Which functional groups in organic matter, act as the most important sites for water adsorption?

Ans. Carboxylic groups

Q.138. What is the optimum Ca:Mg ratio in soil?

Ans. 10 - 15:1

Q.139. What is the optimum ratio of calcium to total cations in soils?

Ans. 1 / 5 to 1 / 20

Q.140. Give examples for 1:1 trioctahedral clay minerals.

Ans. Serpentine and greenalite

Q.141. Give examples for 1:1 dioctahedral clay minerals.

Ans. Nontronite and Beidellite

Q.142. Give examples for 2:1 trioctahedral clay minerals.

Ans. Sauconite, Sepiolite and Hectrotite

Q.143. Give examples for 2:1 dioctahedral clay minerals.

Ans. Pyrophyllite and Polygorskite

Q.144. Name the 2:1 trioctahedral clay mineral that has zircon in its octahedral layer.

Ans. Sauconite

Q.145. Name the 2:1 trioctahedral clay mineral that has nickel ion in its octahedral layer.

Ans. Pimelite

Q.146. Give example for 2:1 clay mineral that has zero electrical charge.

Ans. Talc

Q.147. Name the 1:1 clay mineral with zero electrical charge.

Ans. Serpentine

Q.148. Give examples for 2:1 trioctahedral mica group clay minerals.

Ans. Biotite, phlogobite.

Q.149. Write the formula to calculate gypsum requirement.

Ans. ESP (initial) - ESP (final) / 100 x CEC

Q.150. What is the gypsum equivalent of elemental sulphur?

Ans. 0.18

Q.151. Which secondary clay mineral has high surface area?

Ans. Montmorillonite

Q.152. Name the bonds that join interlayer in vermiculite.

Ans. Magnesium bonding

Q.153. What is the cation exchange capacity of kaolinite clay minerals?

Ans. 3 - 15 Cmol (P+) Kg^{-1}

Q.154. What is the cation exchange capacity of montmorillonite clay minerals?

Ans. 80 - 120 Cmol (P+) kg^{-1}

Q.155. What is the cation exchange capacity of vermiculite clay mineral?

Ans. 120 - 150 Cmol (p+) kg^{-1}

Q.156. Who proposed lignin theory of humus formation?

Ans. Waksman (1932)

Q.157. Who proposed Polyphenol theory of humus formation?

Ans. Stevenson (1982)

Q.158. Who proposed Sugar-amine condensation theory of humus formation?

Ans. Stevenson (1982)

Q.159. In polyphenol theory of humic substances formation, which enzyme convert polyphenols into quinones?

Ans. Phenol oxidases

Q.160. In the soils with poor drainage and sediments (swamps), which theory of humic substances formation will be predominant?

Ans. Lignin theory of humus formation

Q.161. In the soils with dramatic and frequent changes in the environment especially temperature and moisture, which theory of humic substances formation will be predominant?

Ans. Sugar-amine condensation theory of humus formation

Q.162. In the forest soils, which theory of humic substances formation will be predominant?

Ans. Polyphenol theory of humus formation

Q.163. What is the percentage of pH dependent charges on 2:1 clay mineral?

Ans. 5 - 10%

Q.164. What is the percentage of pH dependent charges on 1:1 clay minerals?

Ans. 50%

Q.165. Arrange the soil textural classes according to their CEC (meq / 100g) in decreasing order?

Ans. Clay loam > silt loam > loam > sand

Q.166. What is the root CEC range of monocotyledons?

Ans. 10-15 meq / 100g

Q.167. In the formula, dc / dx=- DA what do you mean by dc / dx?

Ans. Fick's law

Q.168. In ion-carrier hypothesis, mechanism-I is applicable at what concentration of ions?

Ans. Below 1mM

Q.169. In ion-carrier hypothesis, mechanism-II is applicable at what concentration of ions?

Ans. Above 1mM

Q.170. Where can we find Jarosite mottles?

Ans. Sulphuric horizon

Q.171. Which rocks have albite and orthoclase?

Ans. Metamorphic rocks

Q.172. Give an example for basic pyroxene mineral.

Ans. Hyperstene

Q.173. Give examples for amphibole group mineral.

Ans. Hornblende, Actinilite

Q.174. Name a silicate group mineral that has no oxygen sharing for its formation.

Ans. Neosilicates

Q.175. What is the isomorph of Gibbsite mineral?

Ans. Bayerite

Q.176. What is the isomorph of Boehmite mineral?

Ans. Lepidochrosite

Q.177. What is the isomorph of Goethite mineral?

Ans. Marbite

Q.178. Name the Mineral that has structure similar to haematite.

Ans. Martite

Q.179. Write the isomorph of haematite.

Ans. Maghemite

Q.180. Give example for titanium containing mineral.

Ans. Rutile

Q.181. Write the chemical formula of Goethite.

Ans. α - FeOOH

Q.182. Write the chemical formula of Gibbsite.

Ans. α - $Al(OH_3)$

Q.183. Which minerals are formed by the reaction of sulphuric acid with soil minerals?

Ans. Jarosite and Aluminite

Q.184. Write the adsorption equation dealing with multi-layer adsorption in soil.

Ans. BET equation

Q.185. Write the equation used for the heterovalent ion exchange study.

Ans. Gapon's equation, Vanselow equation, KDO equation

Q.186. Name the clay mineral that has high surface charge density.

Ans. Kaolinite

Q.187. What should be the relative pressure of BET adsorption equation?

Ans. 0.05 - 0.45

Q.188. Which is widely used chemical for the calculation of surface area of clay minerals ?

Ans. NH_3

Q.190. Write paulings second rule.

Ans. An ionic structure will be stable to the extent that the sum of the strength of the electrostatic bonds that reach an anion equal the charge on that anion

Q.191. What is the K content in schoenite?

Ans. 22 - 24% K

Q.192. Which form of zinc is dominant in the pH range of 7.0 - 9.0 in soils?

Ans. Zn(OH)+

Q.193. Which form of Zn is dominant in the pH range greater than 7.0 in soils?

Ans. Zn^{2+}

Q.194. Which form of Cu is dominant in the pH range greater than 7.0 in soils?

Ans. $Cu(OH)_2$

Q.195. What does a low phosphate potential indicate?

Ans. High availability of phosphates

Q.196. Give equation of lime potential.

Ans. pH –1 / 2p Ca = Lime potential

Q.197. Which clay minerals have both permanent charge as well as pH dependent charge?

Ans. Montmorillonite, vermiculite, chlorite

Q.198. Who proposed lattice theory of clay mineral?

Ans. Page and Baver

Q.199. What is the Y-index of illite clay mineral?

Ans. 5.0 - 25.0

Q.200. What is the Y- index of montmorillonite clay mineral?

Ans. > 5.0

Q.201. Write the value of C -axis for halloysite clay mineral.

Ans. 10.1°A

Q.202. Write the value of C - axis for illite clay mineral.

Ans. 18°A

Q.203. Write the value of C - axis for chlorite clay mineral.

Ans. 14.0°A

Q.204. Write the value of C - axis for vermiculite clay mineral.

Ans. 14.0°A

Q.206. Write the chemical formula of salt peter.

Ans. KNO_3

Q.207. Write the chemical formula of nitre.

Ans. $Na\ NO_3$

Q.208. Which ions are formed from neutral atoms by gain of electrons?

Ans. Negative ions

Q.209. How many electrons and neutrons does oxygen atom contain?

Ans. 8 neutrons and 8 electrons

Q.210. What is buffering capacity of soil?

Ans. Resistance offered by the soil to change in pH.

Q.211. Which soil texture has the greatest buffering capacity?

Ans. Clay loam

Q.212. Which ion causes more acidity in soil? Why?

Ans. Al^{3+} ions. Because it react with water to produce H^+ ions.

Q.213. What is the redox potential of oxidized zone in submerged soils?

Ans. More than 400 mv

Q.214. In submerged soil, H_2S toxicity occurs in which stage?

Ans. Highly reduced toxic stage.

Q.215. What is the relationship between pE and redox potential.

Ans. $pE = \frac{\text{Redox potential}}{0.0591}$

Q.216. How will be the availability of nutrient ions during submergence in soil?

Ans. * Submergence increases the availability of N, P, Mn, Fe, Mo, Si * Submergence decreases the availability of S, Cu, Zn.

Q.217. Mention the mechanism for increased availability of native and applied P in submerged soils

Ans.
- Submergence increases solubility of strengite and variscite, $Ca\text{-}PO_4$
- Submergence causes reduction of $Fe\text{-}PO_4$

Q.218. State Henderson - Hasselbalch equation

Ans. PA = Pka + log (A)/ (HA)

Q.219. What is the K content of illite?

Ans. 5.0 - 8.0 %

Q.220. Define surface potential.

Ans. Due to the presence of opposite charges on the colloidal surface and in the liquid phase, an electrical potential develops at the solid-liquid interface called as surface potential.

Q.221. What would be the charge of colloid at the pH value above zero point charge?

Ans. Negative

Q.222. Write the methods used to distinguish between expanding and non expanding minerals in X-ray diffraction analysis.

Ans. K Saturation, Mg Saturation, Solvation of Mg saturated samples, K saturation and heating at 500°C

Q.223. Write a non-destructive method of identification of clay minerals

Ans. X-ray diffraction analysis

Q.224. Write the ionic radii of Si^{4+}; $A1^{3+}$; NH^{4+} and K^{+}ions?

Ans. Si^{4+} : 0.039 nm

Al^{3+} : 0.051 nm

NH^{+}_{4} : 0.143 nm

K^{+} : 0.133 nm

Q.225. Write the surface charge density of Kaolinite, Montmorillonite, illite and vermiculite.

Ans. Kaolinite : Zero
Montmorillonite : 0.3 - 0.6
Illite : ~ 2.0
Vermiculite : 0.6 - 0.9

Q.226. Write the surface area of important clay mineral.

Ans. Kaolinite : 7.0 - 30 m^2 / kg
Montmorillonite : 600 - 800 m^2 / kg
Vermiculite : 600 - 800 m^2 / kg
Allophane : 100 - 800 m^2 / kg
Halloysite : 10 - 45 m^2 / kg
Chlorite : 25 - 150 m^2 / kg
Biotite : 40 - 100 m^2 / kg
Muscovite : 60 - 101 m^2 / kg

Q.227. What is the Zeta potential of Li-saturated clay?

Ans. 58.8

Q.228. What is the Zeta potential of NH_4- Saturated clay?

Ans. 56.0

Q.229. What is the Zeta potential of H-Saturated clay?

Ans. 48.4

Q.230. What is the Zeta potential of Ca-saturated clay?

Ans. 516

Q.231. What is pH dependent charge or variable charge?

Ans. The charge resulting from broken chemical bonds on the edges of 1:1 clay is called variable charge or pH dependent charge.

Q.232. What is permanent charge?

Ans. The charge resulting from isomorphous substitution in the structure of 2:1 Clays is permanent charge.

Q.233. What is the size of the cavity formed in between the two layer units of clay minerals in NH_4 + or K^+ fixation?

Ans. 2.8°A

Q.234. Give examples for 2:1 trioctahedral mica group clay minerals.

Ans. Biotite and phlogobite

Q.235. Give examples for 2:1 dioctahedral mica group clay minerals.

Ans. Muscovite, Illite, Glauconite

Q.236. In which 2:1 clay minerals, octahedral substitution is more?

Ans. Montmorillonite

Q.237. In which 2:1 clay mineral tetrahedral substitution is more?

Ans. Nontronite and Bidellite

Q.238. Write the first and second order diffraction at angle of 2θ and d spacing for Kaolinite clay.

Ans. First order diffraction peak at 2θ = 12.4°, d spacing = 7.13°A
Second order diffraction peak at 2θ = 25°, d spacing = 3.56°A

Q.239. Write the first order diffraction at angle of 2θ and d spacing for Montmorillonite clay.

Ans. First order diffraction peak at 2θ = 7.2°, d spacing = 12.3°A

Q.240. Write the first order diffraction peak at angle of 2θ and d spacing for illite clay.

Ans. First order diffraction peak at 2θ = 8.7°, d spacing = 10.1°A

❑❑❑

Chapter-5

FERTILIZERS AND MANURES

Q.1. Give an example for organic chemical fertilizer.

Ans. Urea

Q.2. Which is the basic fertilizer?

Ans. Sodium nitrate

Q.3. Which fertilizer material is partly soluble in water?

Ans. Calcium Ammonium Nitrate (CAN)

Q.4. Which fertilizer is highly hygroscopic?

Ans. Urea

Q.5. What is the percentage of nitrogen in urea ?

Ans. 46%

Q.6. What is the permissible maximum biuret content of urea produced in India ?

Ans. 1.5%

Q.7. How is single superphosphate produced using rock phosphate as raw material?

Ans. By reaction with sulphuric acid

Q.8. How is concentrated super phosphate produced using rock phosphate as raw material?

Ans. By reaction with Phosphoric acid

Q.9. What is the chief component of single super phosphate that supplies phosphorus to plants?

Ans. Mono calcium phosphate

Q.10. Which is the acid soluble phosphate fertilizer?

Ans. Rock phosphate

Q.11. Name the fertilizers containing N in ammoniacal and nitrate form.

Ans. Calcium ammonium nitrate (25% N), ammonium nitrate (33% N) and ammonium sulphate nitrate (26% N).

Q.12. Give examples for nitrification inhibitors.

Ans. N serve [2 chloro - 6 - trichloromethyl Pyridine], Dicynamide, Thiourea, Thiocarbamate.

Q.13. Write the per cent content of N and P in DAP.

Ans. I8 % N, 46 % P_2O_5

Q.14. Which is the water soluble phosphate?

Ans. $Ca(H_2PO_4)_2$ (Mono Calcium Phosphate)

Q.15. What is the conversion factor to convert % P to % P_2O_5?

Ans. 2.29

Q.16. Name the mineral that contains K_2SO_4.

Ans. Langbenite

Q.17. What is the per cent K in KCl?

Ans. 60%

Q.18. Which fertilizer leads to highest Ammonia volatilization loss?

Ans. $(NH_4)_2SO_4$

Q.19. Which fertilizer is physiologically neutral?

Ans. Calcium ammonium nitrate (CAN)

Q.20. What happens when Ammonium sulphate is placed on the surface of alkali soils?

Ans. Volatilization of ammonia occur

Q.21. Which is the highly concentrated widely used solid nitrogenous fertilizer?

Ans. Urea

Q.22. Which nitrogenous fertilizer would you prefer for groundnut grown on light soil?

Ans. Ammonium sulphate

Q.23. Which is the recommendable method of $(NH_4)_2 SO_4$ application to a soil having pH 8-9?

Ans. Surface application with incorporation

Q.24. What is enriched superphosphate?

Ans. Mixture of SSP and FYM is known as Enriched superphosphate.

Q.25. Which fertilizer causes sulphide toxicity in submerged rice soil upon continuous application of high dose?

Ans. $(NH_4)_2 SO_4$

Q.26. Which is the fertilizer that may cause maximum phytotoxicity?

Ans. Urea

Q.27. Which is the fertilizer that is physiologically alkaline?

Ans. $NaNO_3$

Q.28. Which is the fertilizer that induces poor physical condition of soil ?

Ans. $NaNO_3$

Q.29. Which fertilizer induces the leaching loss of Ca in soil?

Ans. $NaNO_3$

Q.30. Which is ammonium forming fertilizer?

Ans. Urea

Q.31. For water logged rice cultivation, which fertilizer should be avoided ?

Ans. $NaNO_3$

Q.32. What is the percentage of phosphorus in organic matter?

Ans. 0.5%

Q.33. What is the percentage of nitrogen in organic matter?

Ans. 5.0%

Q.34. What is the percentage of calcium in single super phosphate?

Ans. 18 - 21%

Q.35. What is the percentage of sulphur in single super phosphate?

Ans. 12%

Q.36. What is the content of P_2O_5 in single super phosphate?

Ans. 16%

Q.37. What is the percentage of P_2O_5 in triple super phosphate?

Ans. 48 - 52%

Q.38. What is the percentage of P_2O_5 in basic slag ?

Ans. 3 - 8%

Q.39. What is the percentage of calcium in basic slag?

Ans. 20%

Q.40. For which soils, single super phosphate is best suitable?

Ans. Neutral and alkaline soils

Q.41. What is the percentage of sulphur in gypsum?

Ans. 21. 0%

Q.42. What is the percentage of S in ammonium sulphate?

Ans. 21.0%

Q.43. What is the percentage of Ca in calcium ammonium nitrate (CAN)?

Ans. 8.0%

Q.44. What is the percentage of Ca in gypsum?

Ans. 30. 0%

Q.45. Which is the first fertilizer produced in India?

Ans. Single super phosphate

Q.46. Which is the high fertilizer consuming state in India?

Ans. Punjab

Q.47. What is percentage of citrate soluble P_2O_5 in rock phosphate?

Ans. 5 - 17%

Q.48. Which phosphatic fertilizer is called as concentrated superphosphate?

Ans. Triple super phosphate

Q.49. What is Utilization Efficiency of fertilizer ?

Ans. The percentage of added fertilizer that is actually used by the plants.

Q.50. The Fertilizer control order (FCO) came into existence to – – – – – – – – – –

Ans. Prevent adulteration in fertilizer manufacture Year of FCO 1957 and amended in 1985.

Q.51. – – – – – – – – – –is an example of urease inhibitor.

Ans. Thiourea

Q.52. What is fertilizer grade?

Ans. Refers to the guaranteed minimum percentage of nitrogen (N), Phosphorus (P_2O_5) and potash (K_2O) contained in the mixed fertilizer.

Q.54. What is fertilizer ratio?

Ans. Ratio of the percentage of nitrogen (N), phosphorus (P_2O_5) and potash (K_2O) in the fertilizer mixture.

Q.55. What are conditioners?

Ans. Materials which are added to fertilizer mixtures during their preparation for reducing hygroscopicity and improve their physical condition. Such materials are low grade organic materials.

Q.56. What per cent of applied P fertilizers is available to plants?

Ans. Less than 20%

Q.57. What is a complete fertilizer?

Ans. Fertilizer that supply all three major nutrients N, P, K.

Q.58. What is straight fertilizer?

Ans. Fertilizer that supply only one major nutrient element.

Q.59. Why Ammonium sulphate is not a suitable fertilizer for upland calcareous soils?

Ans. Due to losses of ammonia through volatilization.

Q.60. Give example for organic N fertilizer.

Ans. Urea (Amide form) $NH_2 - CO - NH_2$

Q.61. Give examples for slow release N fertilizers.

Ans. Coated N fertilizers (Sulphur coated urea, Lac coated urea) N substances of low solubility (Urea Formaldehyde).

Q.62. Classify P fertilizers based on solubility.

Ans. Water soluble fertilizer (Eg. SSP, TSP) Citrate soluble P fertilizer (Eg. Basic slag) Insoluble P fertilizer (Eg. Rock phosphate).

Q.63. What is biosuper?

Ans. Granular material containing mixture of rock phosphate and ground elemental S inoculated with sulphur oxidizing bacteria *(Thiobacillus thioxidans)*.

Q.64. By what processes, muriate of potash is manufactured?

Ans. Floatation and recrystallization processes

Q.65. Name some complex fertilizers

Ans. Nitrophosphate, Polyphosphate

Q.66. Which fertilizer is added to prevent nitrogen losses during the decomposition of organic matter?

Ans. Single super phosphate

Q.67. What is also called as Nitro lime or Nitro - chalk ?

Ans. Calcium ammonium nitrate

Q.68. What is the nitrogen percentage in calcium ammonium nitrate (CAN)?

Ans. 25.0

Q.69. Give an example for explosive fertilizer.

Ans. Ammonium nitrate

Q.70. What is the acid equivalent of 100 kg of ammonium sulphate nitrate?

Ans. 110 kg $CaCO_3$

Q.71. What are the pollutants of soil and water from phosphate fertilizer industries?

Ans. Uranium, Cadmium, Fluoride

Q.72. Define mixed fertilizers.

Ans. The physical mixture of two or more straight fertilizers or compound fertilizer is called mixed fertilizer.

Q.73. What is the temperature maintained in NH_3 production by Haber-Bosch synthesis process?

Ans. 400 - 500°C

Q.74. What is the by-product of Coal distillation?

Ans. Ammonium Sulphate

Q.75. Name the unstable compound first produced during the hydrolysis of urea.

Ans. Ammonium Sulphate

Q.76. Name an ideal N fertilizer suitable for foliar spray.

Ans. Urea

Q.77. Name the fertilizer obtained as a by-product during soda ash manufacture.

Ans. Ammonium Chloride

Q.78. Name some filler materials used in the preparation of mixed fertilizers.

Ans. Sand, Soil, Coal, Organic manures

Q.79. Name some slow release N fertilizers.

Ans. Urea form, Isobutyledene diurea, sulphur coated urea.

Q.81. What is the organic matter content in cattle manure?

Ans. 15.2%

Q.82. What is the N, P and K content (%) in FYM?

Ans. N : 0.5%

P : 0.3%
K : 0.5%

Q.83. What is the average N content in green manures?

Ans. 0.7%

Q.84. Give an example for concentrated organic manure.

Ans. Bone meal, Oil cakes

Q.85. What should be the P concentration in organic matter for net mineralization?

Ans. More than 0.2%

Q.86. Name an anaerobic N_2 fixing bacteria.

Ans. Clostridium

Q.87. Give examples for acid tolerant N_2 fixing bacteria.

Ans. Beijerinckia, Derxia

Q.88. Name some BGA involved in N fixation.

Ans. Anabaena, Aulosira, Anabaenopsis, Cylindrospermum, Nostoc, Tolypothrix.

Q.89. Name the nutrient elements required for N_2 fixation.

Ans. Mo, Fe, Ca, Co

Q.90. Mention the N fixing bacteria associated with legumes.

Ans. Rhizobium

Q.91. Name an aerobic N_2 fixing bacteria.

Ans. Azotobacter

Q.92. Mention the important species of Rhizobium.

Ans.

Rhizobium spp.	Associated Crop / group
R. meliloti	Alfalfa
R. leguminosarum	Peas
R. phaseoli	Beans
K. japonicum	Soybeans
R. lupini	Lupin
R. trifolii	Clover
R. spp	Cowpea

Q.93. Expand FAI.

Ans. Fertilizer Association of India

Q.94. What is fertigation?

Ans. Application of fertilizers through irrigation water.

❑❑❑

Chapter-6

SOIL FERTILITY

Q.1. How many essential elements are required for the growth of most higher plants?

Ans. 17

Q.2. Who is the father of field experiments?

Ans. Jan Baptiste Bousingault

Q.3. Who established water as sole source of nutrient to plants?

Ans. Jan Baptiste Van Helmont through willow tree experiments

Q.4. Which form of nitrogen absorbed by plant is toxic?

Ans. NO_2

Q.5. What is the product of aminization?

Ans. $R\text{-}NH_2$

Q.6. What is the process in which nitrate is reduced ?

Ans. Denitrification

Q.7. By denitrification, which gas largely escapes to the atmosphere?

Ans. N_2O

Q.8. Which organism oxidizes ammonium ions?

Ans. Nitrospira

Q.9. Which organism is leaf-inhabitant?

Ans. Beijerinkia

Q.10. Which micro organism is responsible for nitrification or conversion of nitrite to nitrate?

Ans. Nitrospira, Nitrobacter

Q.9. What is the process of conversion of ammonical nitrogen to nitrate nitrogen? Which organism is responsible for conversion?

Ans. Nitrification, Nitrosomonas

Q.10. In which soils, volatilization loss of nitrogen is more ?

Ans. Alkaline soils

Q.11. Which enzyme converts amide form of nitrogen to ammonium carbonate when urea is added to soil?

Ans. Urease

Q.12. Mention the most important factor influencing nitrification in soil.

Ans. Soil aeration

Q.13. Rank the soils in the order of ammonia volatilization capacity?

Ans. Calcareous soil > Alkaline soil > Acidic soil

Q.14. What is the form of nitrogen that is specific to denitrification?

Ans. Nitrate

Q.15. Which form of P is more available to plants?

Ans. Mono calcium phosphate in super phosphate

Q.16. What is the predominant ionic species of phosphate in soil at pH below 6.0?

Ans. H_2PO_4

Q.17. What is the predominant phosphate in single super-phosphate?

Ans. Mono calcium phosphate

Q.18. Which nutrient stimulates root development?

Ans. Phosphorus

Q.19. In which soil, fixation of phosphate is more?

Ans. Acid soils

Q.20. In anaerobic soils, which organism causes sulphate reduction?

Ans. Desulfotomaculum

Q.21. What is the soil pH in which Nitrites largely accumulates?

Ans. pH 8.0

Q.22. What is the range of total N content in Indian soils?

Ans. 0.02 - 2.5%

Q.23. What is the range of total potassium content in Indian soils?

Ans. 0.05 – 3.5%

Q.24. What is the range of total phosphorus content in Indian soils?

Ans. 0.02 – 0.15%

Q.25. How many times does the organic matter hold its weight of water?

Ans. Up to 20 times

Q.26. What is the percentage of carbon utilized by the organisms in soil?

Ans. 35%

Q.27. At what C:N ratio, does the immobilization of soil nitrogen occurs ?

Ans. > 30 : 1

Q.28. In which C : N ratio of organic matter does mineralization of soil nitrogen occur?

Ans. < 20:1

Q.29. When organic matter is incorporated in soil, at which C:P ratio, mineralization of soil phosphorus occurs?

Ans. < 200:1

Q.30. What is the desired ratio of NPK use?

Ans. 4:2:1

Q.31. What is the commonly used NPK ratio in India?

Ans. 8.9:2.7:1

Q.32. Who first established that plants obtain their carbon supply from CO_2 in air and from the soil?

Ans. Justus Von Liebig

Q.33. Which nutrient element is the basic constituent of life?

Ans. Nitrogen

Q.34. Which nutrient is essential for the increase in protein content in food and fodder crops?

Ans. Nitrogen

Q.35. Which nutrient brings about the early maturity of crops, particularly the cereals, and counteracts the effects of excessive nitrogen?

Ans. Phosphorus

Q.36. Which nutrients stimulates flowering and aids in seed formation ?

Ans. Phosphorus

Q.37. Which nutrient, when applied to legumes, enhances the activity of rhizobia and increase the formation of root nodules?

Ans. Phosphorus

Q.38. Which element, when present in excess quantity may cause deficiencies of trace elements, particularly iron and zinc?

Ans. Phosphorus

Q.39. Which element is not a constituent of plant composition?

Ans. Phosphorus

Q.40. Which element increases the vigour and disease resistance to plants?

Ans. Potassium

Q.41. Which nutrient increases the winter hardiness and reduces lodging of crops?

Ans. Potassium

Q.42. What is the purpose of nitrification inhibitors?

Ans. To stop rapid mineralization of NH_4^+ ions, to reduce leaching losses of N to ground water

Q.43. How can the availability of P to plants from rock phosphate be improved?

Ans. Premixing it and composting it with organic amendments prior to application

Q.44. In consideration of plant availability of all essential nutrients, what is a typical range for the optimal pH for many agricultural soils?

Ans. 6.0 - 7.5

Q.45. Which combination of solid phases of phosphate compounds would be typically found in acid and alkaline soils?

Ans. Fe and Al phosphates in acid soil and tricalcium phosphate in alkaline soils.

Q.46. How can decomposition of organic matter cause soil acidity?

Ans. Releases CO_2 which forms H_2CO_3

Q.47. Which nutrient is an essential constituent of certain volatile compounds and helps in the reduction - oxidation system in respiration?

Ans. Sulphur

Q.48. Which element acts as a regulator of potassium / calcium ratio in the plant?

Ans. Boron

Q.49. Of the micronutrients, which element is essential for translocation of sugars in plant?

Ans. Boron

Q.50. Of the micronutrients, which element is essential for flowering, fruiting process and pollen germination?

Ans. Boron

Q.51. Which nutrient element influences the formation of some growth hormones in the plant?

Ans. Zinc

Q.52. Which element is essential for nitrogen fixation, both symbiotic and non symbiotic ?

Ans. Molybdenum

Q.53. Which element acts as "electron carrier" in enzymes which brings about oxidation-reduction reaction in plants?

Ans. Copper

Q.54. Which is the element essential for Ca metabolism, both Ca uptake by roots and its efficient use in plants ?

Ans. Boron

Q.55. Which element helps in utilization of iron in chlorophyll synthesis?

Ans. Copper

Q.56. Which is the largest loss of nutrients from the soil?

Ans. Removal by crops

Q.57. What metal element is related to siderophores?

Ans. Iron

Q.58. What is the CN ratio of organic matter?

Ans. 10:1

Q.59. Which reaction is essential for keeping a micronutrient in solution and available to plants by forming an organic complex around the micronutrient?

Ans. Chelation

Q.60. Which conditions cause Zn deficiency?

Ans. High pH (alkaline under anaerobic conditions)

Q.61. Which soil is prone to Zn deficiency?

Ans. Calcareous soils containing a large amount of bicarbonate.

Q.62. Which nutrient deficiency causes Bitter pit in apple, leaf tip burn in cabbage and lettuce, black heart of celery, Blossom end rot of tomato, cavity spot of carrots?

Ans. Calcium

Q.63. Which nutrient deficiency causes 'Khaira' disease in rice?

Ans. Zinc

Q.64. Which nutrient deficiency results in the symptoms of "little leaf" or "rosetting" and shortening of internode?

Ans. Zinc

Q.65. Which nutrient deficiency results in 'Brown heart' in radish and crown choking in coconut?

Ans. Boron

Q.66. Which nutrient deficiency causes 'Whiptail' in cauliflower?

Ans. Molybdenum

Q.67. Which condition indicates Fe / Mn toxicity in rice?

Ans. Reddish brown coloration of water in rice field

Q.68. Yellowing of young leaves is due to – – – – – – – deficiency

Ans. Sulphur

Q.69. Which is the most important deficiency symptom of nitrogen in cereals?

Ans. Severe restriction of tillering with stem dying and bronzing

Q.70. Which is the most common symptom(s) of K deficiency?

Ans. Leaf tips and margins become scorched and necrotic

Q.71. Which is the most common symptom(s) of Ca deficiency?

Ans. Retarded growth of both shoots and roots

Q.72. Which is the critical symptom(s) of boron deficiency in cauliflower?

Ans. Young curds, developing small water soaked areas

Q.73. Which nutrient deficiency is similar to visual symptoms of sulphur deficiency?

Ans. Nitrogen

Q.74. Which nutrient has negligible residual effect? Why?

Ans. Nitrogen, Because it is highly mobile and lost by leaching

Q.75. What is the order in which Ammonia volatilization from different N fertilizer sources increases?

Ans. Ammonium sulphate $<$ Urea $<$ ammonium bicarbonate

Q.76. What is Chlorosis?

Ans. Yellowing of leaves

Q.77. What is Necrosis?

Ans. Complete drying and death of plant tissue

Q.78. What causes internal cork of apple?

Ans. Boron deficiency

Q.79. What causes stem melanosis disease of wheat?

Ans. Copper deficiency

Q.80. What causes bronzing in rice?

Ans. Fe toxicity

Q.81. What causes crinkle leaf of cotton?

Ans. Toxicity of Mn

Q.82. What is the nutrient element involved in the synthesis of tryptophan?

Ans. Zinc

Q.83. What causes fern leaf in potato?

Ans. Deficiency of Zn

Q.84. What causes little leaf of cotton?

Ans. Deficiency of Zn

Q.85. What causes speckled yellow of sugar beet?

Ans. Deficiency of Mn

Q.86. What causes white bud in maize / sorghum?

Ans. Zn deficiency

Q.87. What causes grey speck of oats and marsh spot of peas?

Ans. Mn deficiency

Q.88. Which element is an important component of urease enzyme?

Ans. Nickel

Q.89. Name the indicator plant used for the deficiency of boron.

Ans. Sunflower or Sugar beet

Q.90. Name the indicator plant used for the deficiency of sodium.

Ans. Sugar beet

Q.91. What is Eh in well oxidized soil at pH 7.0?

Ans. +400 mV to + 700 mV

Q.92. If soil Eh at pH 7.0 varies from - 300 mV to - 100 mV what is the soil condition?

Ans. Highly reduced

Q.93. Which nutrients are highly mobile in plants?

Ans. N, P, K

Q.94. Which nutrients are moderately mobile in plants?

Ans. Mg, S, Mo

Q.95. Which nutrients are immobile in plants?

Ans. Ca, Zn, B

Q.96. What nutrient deficiency causes interveinal chlorosis first on older leaves?

Ans. Magnesium

Q.97. Mention Mitscherlich laws.

Ans. Mitscherich's Law of Physiological relationship, Mitscherlich's Growth law

Q.98. Give Baule units of N, P, K

Ans. 1 Baule of N = 223 lbs of N, 1 Baule of P - 45 lbs of P_2O_5, 1 Baule of K = 76 lbs of K_2O

Q.99. Who proposed Criteria of essentiality of elements?

Ans. Arnon and Stout (1939)

Q.100. Give the mechanisms of nutrient uptake.

Ans. Mass flow, Diffusion, Root Interception

Q.101. What are the methods of transport of ions into the plant system?

Ans. Active transport and passive transport

Q.102. What would be the C:N ratio of organic materials for mineralization and immobilization?

Ans. For mineralization < 20:1

For immobilization >30 : 1

Q.103. What would be the C:N:P ratio of organic materials for mineralization ?

Ans. C : N : P = 100:10:1

Q.104. What should be the C:P ratio for mineralization of organic materials ?

Ans. < 200:1

Q.105. What are the mechanisms of P fixation in soils?

Ans. Sorption and precipitation reactions

Q.106. Define phosphate potential.

Ans. Phosphate potential = 1/2 PCa + PH_2PO_4 Higher the phosphate potential, lower is the phosphate concentration.

Q.107. Name some K bearing Minerals.

Ans. K feldspar (Orthoclase, Microcline $KAl\,Si_3O_8$, Muscovite [$KAl_3\,Si_3O_{10}\,(OH)_2$], Biotite [$K\,(Mg, Fe)_3\,Al\,Si_3O_{10}\,(OH)_2$].

Q.108. Who developed Q / I Concept?

Ans. Beckett

Q.109. What is the use of Q / I ratio?

Ans. Used to predict the status of K in soils.

Q.110. What is potential buffering capacity of K (PBC)?

Ans. Measure of ability of a soil to maintain K concentration in the soil solution.

Q.111. What are the forms of Ca in soils?

Ans. Primary Mineral (Anorthite, Basalt), Nodular form ($CaCO_3$), Salt ($CaCl_2$, $CaSO_4$,) exchangeable Ca in soil exchange complex.

Q.112. What is the range of Ca Content of Plants?

Ans. 0.1 to 5.0%

Q.113. What is hypomagnesemia ?

Ans. Mg deficiency in forage crops lead to abnormal low level of Mg in blood, causing a disease called hypomagnesemia in cattle.

Q.114. What should be the C:S ratio required for mineralization of organic materials ?

Ans. C : S ratio < 200:1

Q.115. What are different types of manures?

Ans. Bulky organic manure (Eg. FYM, Compost), Concentrated Organic manure (Eg. Oil cake, Bone meal)

Q.116. Write the kinds of green manuring.

Ans. Green manures Ex. Sunhemp *(Crotolaria, juncea)*, Daincha (Sesbania aculeata)
Green leaf manures Ex. Glyricidia *(Glyricidia maculata)*

Q.117. What is active absorption of ions?

Ans. Movement of ions in to plant roots against an activitiy gradient. This process requires energy.

Q.118. Who proposed the law of minimum?

Ans. Justus Von Liebig (1840)

Q.119. What is Luxury consumption of nutrients?

Ans. Amount of nutrients absorbed by plants in excess of their requirement for optimum growth.

Q.120. What is mass flow?

Ans. Nutrient ions transported in flow of water due to transpiration, evaporation and percolation of soil water.

Q.121. What is neutralizing index (NI) of liming material ?

Ans. NI = Calcium carbonate equivalent x fineness factor

Q.122. What is N factor?

Ans. Number of units of organic N immobilized for each 100 units of decomposing organic material.

Q.123. What should be average value for N factor ?

Ans. 0.1 - 1.3

Q.124. Mention important functions of N.

Ans. Impart vegetative growth and it is an integral part of chlorophyll.

Q.125. Mention important functions of P.

Ans. Energy storage and transfer, constituent of nucleic acid, phytin and phospholipids, stimulate root growth, seed formation.

Q.126. Mention important functions of K.

Ans. Enzyme activation, maintain water balance and impart vigour and disease resistance to plants.

Q.127. What is hidden hunger?

Ans. Reduction in yield without the appearance of deficiency symptoms.

Q.128. Mention important functions of Ca.

Ans. Constituent of cell wall and has essential role in cell elongation and division.

Q.129. Mention important functions of Mg.

Ans. Constituent of chlorophyll, cofactor of enzymes

Q.130. Mention important functions of S.

Ans. Required for synthesis of S containing amino acids viz., cystine,cysteine, methionine, formation of chlorophyll, activate proteolytic enzymes.

Q.131. Mention important function of Fe.

Ans. Formation of chlorophyll, structural component of porphyrin molecules like cytochrome, ferrichrome leghaemoglobin.

Q.132. Mention important functions of Mn.

Ans. Chlorophyll formation, takes part in oxidation - reduction, decarboxylation and hydrolysis reactions, influence auxin levels.

Q.133. Mention important functions of Cu.

Ans. Forms various compounds with amino acids and proteins, act as electron carrier in enzyme.

Q.134. Mention important functions of boron.

Ans. Involve in Ca metabolism, regulator of K / Ca in higher plants necessary for pollination, fruit setting, translocation of sugar, formation of nodules

Q.135. Mention important functions of molybdenum.

Ans. Component of nitrate reductase, nitrogenase, important for Fe absorption and translocation.

Q.136. Who classified the elements based on the role of element in plant system.

Ans. Truog, 1954
Structural elements - C, H, O
Accessory Structural elements - N, P, S
Regulator and carriers - K, Ca, Mg
Catalyst and Enzyme activators - Fe, Mn, Zn, Cu, Mo, Cl, B

Q.137. Where was the first permanent manurial experiment in India started?

Ans. Kanpur

Q.138. What is the first unstable compound produced by the hydrolysis of urea?

Ans. Ammonium carbamate

Q.139. Give examples of natural chelates.

Ans. Citric acid and oxalic acid

Q.140. What is the method used to measure a nutrient concentration in soil at equilibrium with the same in soil solution?

Ans. E - value

Q.141. Which disease is caused by nitrate pollution?

Ans. Methemoglobinaemia

Q.142. Which fertilizer contains nutrients other than primary nutrients?

Ans. Urea

Q.143. What is KCl. $MgCl_2$. 6 H_2O?

Ans. Carnalite

Q.144. Who discovered the essentiality of iron ?

Ans. Gris

Q.145. Which nutrient is essential for stomata regulation in plants?

Ans. Potassium

Q.146. How are Biofertilizers applied ?

Ans. Seed treatment, Seedling root dip, Soil treatment

Q.147. Mention the two crop growth factors.

Ans. Genetic and Environmental factors.

Q.148. What is the range of temperature for optimum growth of the agricultural crops?

Ans. 15°C to 40°C

Q.149. Who proposed the law of physiological relationships and growth law?

Ans. Mitscherlich (1909)

Q.150. What is the amount of nutrients required to produce yield that is 50 percent of maximum possible yield?

Ans. One Baule unit

Q.151. Who proposed the nutrient mobility concept?

Ans. Bray

Q.152. Define soil Productivity.

Ans. The capability of soil to produce good crop is called as soil productivity.

Q.153. Define soil fertility.

Ans. The inherent capacity of soil to provide nutrients is called as soil fertility.

Q.154. Who defined the functional metabolism of nutrients?

Ans. D.J. Nicholas Eg. Si, Na, Co, Se

Q.155. At what pH, phosphate ions in solution is precipitated by hydrated Fe?

Ans. pH 3.0 - 7.0

Q.156. At what pH, phosphate ions in solution is precipitated by hydrated Al?

Ans. pH 3.5 - 9.0

Q.157. At what pH, the availability of Ca and Mg is low?

Ans. pH more than 8.5

Q.158. At what pH, the availability of Mo in soil is more?

Ans. pH more than 6.5

Q.159. Who proposed the law of Diminishing Return?

Ans. Mitscherlich.

Q.160. State Mitscherlich equation

Ans. $\log (A-Y) = \log A - Cx$

Q.161. In Mitscherlich equation, what is the efficiency factor C for N, P and K.

Ans. N : 0.122, P : 0.60, K : 0.40

Q.162. State Spillman's equation

Ans. $Y = M (1-R^x)$

Q.163. Who was the first scientist to quantify the relationship between plant growth and addition of a growth factor?

Ans. Mitscherlich

Q.164. Who was the first scientist to initiate the nutrient diagnosis in soil by pot culture methods?

Ans. Mitscherlich

Q.165. Who was the first scientist to initiate the nutrient diagnosis in soil by Aspergillus niger method?

Ans. Mehlich

Q.166. Who was the first scientist to initiate the nutrient diagnosis by plaque method?

Ans. Sackett and Stewart

Q.167. What is the value of low nutrient index?

Ans. < 1.5

Q.168. What is the value of high nutrient index?

Ans. > 2.5

Q.169. What is the Nitrogen use efficiency for rice crop?

Ans. 28 - 34%

Q.170. What is the NUE for crops other than rice?

Ans. 40 - 60%

Q.171. Name the sources of hydrogen ions in soils.

Ans. Carbonic acid, acids released from microbial activity and decomposition of organic matter.

Q.172. Name a highly resistant organic compound.

Ans. Phytin

Q.173. What is the Mg content in plant Chlorophyll?

Ans. 15 - 20%

Q.174. What is the Phosphorus use efficiency in soil?

Ans. 10 - 20%

Q.175. Name the carrier for cation in active transport of nutrient in plant system.

Ans. Nucleic acid

Q.176. Name the carrier for anion in active transport of nutrient in plant system.

Ans. Protein

Q.177. How many number of ATP molecules are required for reduction of one mole of NO_3 to NH_3?

Ans. 15

Q.178. How many number of ATP molecules are required for assimilation of one mole of NH_3?

Ans. 5

Q.179. Name the biological method for evaluation of available Cu and Mg.

Ans. Mulder's Aspergillus Niger method

Q.180. Name the biological method for evaluation of available P and K.

Ans. Mitcherlich pot culture method

Q.181. Name the biological method for evaluation of available micronutrients.

Ans. Neubauer seedling technique

Q.182. What is the Nutrient Use Efficiency of K?

Ans. > 60%

Q.183. What is the Nutrient Use Efficiency of Micronutrients?

Ans. 3 - 5%

Q.184. What are sources of plant nutrients C, H and O?

Ans. Air and water.

Q.185. Name the nutrient available for plant uptake by air pollution?

Ans. S

Q.186. State the concentration of nutrient elements in plants.

Ans. N, P, K, Ca : > 0.1%
Fe, Mn, Zn, Cu : < 0.1%

Q.187. Name the scientist who established the concept of tillage.

Ans. Jethro Tull

Q.188. Who established Rothamstead Experiment station and demonstrated the essentiality?

Ans. J.B Lawes and J.H. Gilbert (1843)

Q.189. What is steenberg effect?

Ans. Under severe nutrient deficiency, rapid increase in yield with added nutrients decrease the nutrient concentration (dilution effect)

❑❑❑

Chapter-7

CHEMISTRY OF CROPS

Q.1. Give the names of protein of cereals.

Ans. Water soluble proteins (albumins), Salt soluble proteins (Globulins), Alcohol soluble proteins (Prolamines), Acids and alkali soluble proteins (Glutelins).

Q.2. Which is responsible for the elastic and cohesive properties of wheat flour?

Ans. Gluten

Q.3. What is the name of rice protein?

Ans. Oryzenin

Q.4. What is the name of protein in red gram?

Ans. Cajanin and Cancajanin

Q.5. What are the active principles in sesamum oil?

Ans. Sesamin and sesamolin

Q.6. Which volatile oil is responsible for the pungency in onion?

Ans. Allyl propyl disulphide

Q.7. What is the name of prolamine in maize?

Ans. Zein

Q.8. What is alkali hydrolysis of fats and oils?

Ans. Saponification

Q.9. Which is a measure of degree of unstauration of fats and oils?

Ans. Iodine value

Q.10. Which is the most important cereal enzyme?

Ans. Amylase

Q.11. What is the toxic substance isolated from polished rice?

Ans. Isolecithin $C_{24}H_{50}NPO_7$

Q.12. Expand NPU, How is it calculated ?

Ans. Net Protein utilization

$$NPU = \frac{\text{True digestibility (TD)}}{100} \times \text{Biological value (BV)}$$

Q.13. What are the pigments present in coloured rice varieties?

Ans. Cyanidin and Delphinidin

Q.14. Which is the scented rice?

Ans. Basmati

Q.15. What is the pretreatment process before milling of rice?

Ans. Parboiling

Q.16. What is the antioxidant present in rice bran oil?

Ans. Tocopherol

Q.17. What is the characteristic protein of Ragi?

Ans. Eleusinin

Q.18. What is the Cyanogenetic glucoside formed on germination of sorghum?

Ans. Dhurrin

Q.19. What is the prolamine present in sorghum?

Ans. Kafirin

Q.20. What is the disease caused due to high leucine concentration in sorghum?

Ans. Pellagra

Q.21. What is responsible for low palatability and digestibility of sorghum food products?

Ans. Tannins

Q.22. What is the name of alkaloid present in sorghum?

Ans. Hordenine (C_{10} $H_{15}ON$)

Q.23. What are the major carotenoids in sorghum?

Ans. Lutein, Zeaxanthine and Beta carotene

Q.24. What is the process of hydrolysis of starch when cereal germinates?

Ans. Malting

Q.25. What is the reaction involving oxidation of water in photosynthesis?

Ans. Light reaction

Q.26. What is the process of reduction of CO to form carbohydrates?

Ans. Dark reaction.

Q.27. Which is the most important site of storage of starch in cereals?

Ans. Seed

Q.28. What are the components of starch?

Ans. Amylose and Amylo pectin

Q.29. Which is the major protein in pulses?

Ans. Globulin

Q.30. What are the two ways for synthesis of amino acids?

Ans. a) Reductive amination b) Transamination

Q.31. How are the amino acids required for protein synthesis obtained?

Ans. By the process of transamination of glutamic and aspartic acids.

Q.32. What are triglycerides?

Ans. Esters of glycerol with fatty acid

Q.33. Give an example for trihydric alcohol.

Ans. Glycerol

Q.34. Give examples of saturated fatty acids.

Ans. Palmitic acid, Stearic acid, Butyric acid, Arachidic acid.

Q.35. Give examples of unsaturated fatty acids.

Ans. Oleic acid, Linoleic acid, Linolenic acid, Ricinoleic acid.

Q.36. What is the globulin present in Groundnut?

Ans. Arachin and Conarachin

Q.37. Give an example for non drying oil.

Ans. Ground nut oil, Olive oil

Q.38. What is the textile fibre manufactured from Groundnut protein?

Ans. Ardil

Q.39. Which is responsible for the good keeping quality of sesame oil?

Ans. Sesamolin

Q.40. What are the tests to find out adulteration of other oil with sesame oil?

Ans. Boudouin test and Villavecchia test

Q.41. Which is the predominant saturated fatty acid in coconut oil?

Ans. Lauric acid

Q.42. Why does coconut oil become dry in winter season?

Ans. low iodine value since, the unsaturated fatty acids are less (< 9%).

Q.43. Which is responsible for the characteristic odour of coconut oil.

Ans. d-lactone and n-octanol

Q.44. Which compound is responsible for the awful odour of coconut oil during deterioration?

Ans. methyl ketone of d-lactones' and n - octanol

Q.45. What are the toxic principles in castor seeds?

Ans. a) Ricin b) Ricinine and c) Allergen

Q.46. Which is the predominant unsaturated fatty acid in castor oil?

Ans. Ricinoleic acid.

Q.47. Give example for drying oil.

Ans. Linseed oil, Safflower oil

Q.48. Give examples for semidrying oil.

Ans. Coconut oil, Cotton seed oil.

Q.49. When oil is good for consumption?

Ans. Before heating

Q.50. What is the colouring matter in safflower?

Ans. Carthamin ($C_{21}H_{22}O_{11}$)

Q.51. What are the steps in synthesis of fats and oil?

Ans. Synthesis of glycerol, Synthesis of fatty acids, condensation of glycerol and fatty acids.

Q.52. What are the important sugars in sugarcane?

Ans. Sucrose, glucose and fructose

Q.53. Which enzyme is responsible for slow inversion of sucrose after harvesting?

Ans. Invertase.

Q.54. Which glucoside is responsible for the odour in Banana?

Ans. Amyl acetate

Q.55. Which carboxylic acid is found in apples ?

Ans. Malic acid

Q.56. Which carboxylic acid is found in Green apples?

Ans. Fumaric acid

Q.57. Which Carboxylic acid is found in Grapes?

Ans. Tartaric acid

Q.58. Give examples of aromatic acids in apple and Banana.

Ans. Quinic acid, shikimic acid

Q.59. Which are the chief amino acids in fruits?

Ans. Aspartic and glutamic acids

Q.60. Which volatile compound is responsible for odour in apple?

Ans. Ethyl 2 methyl butyrate

Q.61. Which volatile compound is responsible for odour in Grapes?

Ans. Methyl anthranilate

Q.62. Which volatile compound is responsible for odour in Orange?

Ans. Limonene

Q.63. What are major flavanoids in fruits?

Ans. Anthocyanin, Leuco anthocyanin, flavones and flavonols.

Q.64. What is the main pigment in Brinjal?

Ans. Anthocyanin known as Nasunin

Q.65. What is the bitter principle of brinjal leaves and fruits?

Ans. Solasodine

Q.66. What is the principal protein of Potato?

Ans. Tuberin

Q.67. What is the name of steroidal glycoalkaloid of potato?

Ans. Solanine

Q.68. What is poor man's orange?

Ans. Tomato

Q.69. What are the flavonol pigments in Bhendi?

Ans. Gossypectin, Quercetin

Q.70. Which is the source of vitamin A?

Ans. Carrot

Q.70. What is the main component of lemon grass oil?

Ans. Citral

Q.72. Which is the ester present in oil of geranium?

Ans. Geraniol

Q.73. Which is the cis isomer of geraniol?

Ans. Nerol

Q.74. Which is the main constituent of pepper mint oil?

Ans. Menthol

Q.75. Which is the chief constituent of oil of turpentine?

Ans. Pinene

Q.76. What is the name of the non volatile pungent oil obtained from Ginger?

Ans. Gingerol

Q.77. Which is the chief sesquiterpene in ginger?

Ans. Zingiberene

Q.78. Which is the major pigment in turmeric?

Ans. Curcumin

Q.79. Which is the major component in turmeric oil?

Ans. Turmerol

Q.80. Which is the most abundant glucoside in vanilla?

Ans. Glucovanillin

Q.81. What is the major constituent of coriander?

Ans. Coriandrol

Q.82. What is the pungent principle of black pepper?

Ans. Piperine

Q.83. What is the major constituent of cinnamon bark oil?

Ans. Cinnamaldehyde

Q.84. What is the chief phenol in clove?

Ans. Eugenol

Q.85. What is the pungent principle in chillies?

Ans. Capsicin ($C_{18}H_{27}O_3N$)

Q.86. Which toxic principle in forage causes infertility in animals?

Ans. Oestrogen

Q.87. Which is the product obtained from cassava starch?

Ans. Sago

Q.88. What is the predominant mineral in sweet potato?

Ans. Potassium.

Q.89. Which is the toxic alkaloid in dioscorea?

Ans. Dioscorin

Q.90. What is the saponin produced by yam?

Ans. Sapogenin

Q.91. What are the sapogenins isolated from dioscorea?

Ans. Diosgenin, botogenin and kryptogenin

Q.92. What is the phospholipid in soybean?

Ans. Lecithin

Q.93. Which is the alkaloid present in tobacco?

Ans. Nicotine, Nornicotine, Neonicotine

Q.94. Give examples for purine group of alkaloids.

Ans. Caffeine, Theobromine, Theophylline.

Q.95. Which is important alkaloid in Glory lily?

Ans. Colchicine

Q.96. Which are the chief alkaloids of opium?

Ans. Morphine, Codeine, Papaverine, Narcotine, Narceine, Thebaine

Q.97. Mention the cynaphoric glucoside in bitter almond?

Ans. Amygdalin.

Q.98. Give an example for Mustard oil.

Ans. Sinigrin.

Q.99. Give examples for phenol glucosides.

Ans. Baptisin in roots of wild indigo, Phloridzin in bark of apple, pear, plum.

Q.100. What is coniferin?

Ans. Alcohol glucoside found in bark of fir tree.

Q.101. Give examples for alcohol glucosides.

Ans. Populin, Salicin, Coniferin.

Q.102. Give examples of dyes obtained from glucosides.

Ans. Indigo, alizarin red

Q.103. Give examples of enzymes which hydrolyze glucosides.

Ans. Myrosin, Emulsin

Q.104. Which is responsible for brown colour of many of cut fruits?

Ans. Oxidation of tannins.

Q.105. What are the broad groups of tannins?

Ans. Pyrogallol tannins, Catechol tannins

Q.106. What are the types of plant pigments?

Ans. Plastid pigments and cell sap pigments

Q.107. Give examples for plastid pigments.

Ans. Chlorophyll and Carotenoid

Q.108. Give examples for cell sap pigments.

Ans. Anthocyanin and Anthoxanthin

Q.109. Which is the precursor of vitamin A?

Ans. Beta Carotene

Q.110. Which is the carotenoid in tomato?

Ans. Lycopene

Q.111. Which is the pigment used for pH indication?

Ans. Anthocyanidin

Q.112. What is PUFA & MUFA?

Ans. Poly unsaturated fatty acid and mono unsaturated fatty acid

Q.113. What is EFA?

Ans. Essential Fatty Acids

❑❑❑

Chapter-8

RADIOISOTOPES

Q.1. Which experiment demonstrated the existence of the particles in the nucleus of an atom?

Ans. Alpha ray scattering experiment (Gold foil experiment by Ernest Rutherford)

Q.2. What spectrum will be formed due to transition of an electron from M level to K level?

Ans. Emission spectra

Q.3. An element X has mass number 32 and an atomic number 16. What is the symbol of its ion?

Ans. X-2

Q.4. For a hydrogen atom En = - 13.6/n2Ev/atom. Therefore minimum energy that an atom of hydrogen in the ground state can absorb is?

Ans. 13.6 eV

Q.5. Who discovered Natural radioactivity?

Ans. Henri Becquerel

Q.6. How isobars are produced ?

Ans. By emission of beta rays by an atom.

Q.7. What does the Nucleus consist of?

Ans. Protons and neutrons

Q.8. Which is the heaviest naturally occurring atoms?

Ans. Uranium

Q.9. Curie is the quantity of radioactive isotope which decays at the rate of.................. disintegrations per sec.

Ans. 3.7 x 1010

Q.10. Gamma rays are not regarded as particles because they have ....

Ans. No mass

Q.11. Gamma rays are not reflected by strong magnetic field because........................

Ans. They carry no charge

Q.12. Name some natural radioactive isotopes.

Ans. Po - 84, Ra - 88, U - 92

Q.13. The alpha particles are nuclei of helium atom and consists of ....

Ans. Two neutrons and two protons

Q.14. Mention the radio active elements discovered by Marie and Pierre.

Ans. Radium and Polonium

Q.15. What is the SI unit of radio activity?

Ans. Becquerel (Bq) 1 Bq - one disintegration / second

Q.16. Define mass number.

Ans. Sum of total of protons and neutrons in the nucleus A=Z+N

Q.17. Why alpha (a) and beta (b) radiations are known as corpuscular radiations?

Ans. Because they are particles in nature.

Q.18. Mention some common ionising radiations.

Ans. These include electromagnetic high energy radiations such as x rays and γ-rays and corpuscular high energy radiations such as electrons, neutrons, protons, deutrons and alpha particles etc..

Q.19. What is RAD?

Ans. Radiation Absorbed Dose. It is the measure of the amount of energy absorbed by the material subjected to radiation

Q.20. Gray unit is equivalent to..........................

Ans. 100 rads

Q.21. Define Atomic number.

Ans. Number of protons in an atom usally denoted by Z

Q.22. Define Isotopes.

Ans. Atoms of the elements having same atomic number but different mass numbers.

Q.23. Define Isobars.

Ans. Atoms of the elements having same mass number but different atomic number.

Q.24. Define Isotones.

Ans. Atoms of elements having different atomic and mass numbers but same number of neutrons.

Q.25. Define Abundance ratio.

Ans. Concentration of an isotope in a naturally occuring element relative to the total content of the element.

Q.26. Define Carrier.

Ans. A stable isotope of a natural element to which radioactive atoms of the same element can be added to obtain a quantity of radioactive mixture sufficient for handling e.g. stable H3PO4 is used as a carrier for H3 32PO4.

Q.27. Define Tracer.

Ans. An isotope of an element, a small amount of which may be incorporated into a sample of material (the carrier) in order to follow (trace) the course of that element through a chemical, biological or physical process and thus also follow the large sample.

Q.28. Define Half life of radionuclides.

Ans. The time in which half the atoms of particular radioactive substance disintegrates to another nuclear form is called the half life.

Q.29. Define Effective half life.

Ans. The time required for the amount of a radioactive element deposited in a living organism to be diminished by 50% as a result of the combined action of radioactive decay and biological elimination.

Q.30. Name the SI Unit of absorbed dose.

Ans. Gray, (Gy) (J/Kg)

Q.31. The spontaneous emission of radiation from the nucleus of unstable isotope is called......................

Ans. Radioactivity

Q.32. What is radioisotope?

Ans. An unstable isotope of an element that decays or disintegrates spontaneously emitting radiation. Approximately 5000 natural and radioactive isotopes have been identified.

Q.33. Name some radio isotopes useful in soil / plant studies.

Ans. ^{32}P, ^{33}P, ^{45}Ca, ^{35}S, ^{59}Fe, ^{65}Zn

Q.34. What is Ndff?

Ans. Fraction of N in the plant derived from the ^{15}N labelled fertiliser. (atom % I5N excess plant sample)
% Ndff = ... x 100 (atom % I5N excess fertilizer)

Q.35. Name the radioisotopes commonly used in root activity studies.

Ans. ^{32}P and ^{86}Rb.

Q.36. The neutron method of measuring the soil water uses the principle of ....

Ans. Neutron thermalisation.

Q.37. Name the Scientist who first used the radioactive tracer technique in biology.

Ans. George Von Hevesy.

Q.38. How is the age of the ancient artifacts determined using the radioisotopes?

Ans. By radio carbon dating

Q.39. Name the scientist who introduced the concept of isotope?

Ans. F. Soddy

Q.40. Name the instrument that uses the slight differences in mass to separate different isotopes of the same element.

Ans. Mass spectrometer

Q.41. What is curie?

Ans. The basic unit used to describe the intensity of radioactivity in a sample of material 1 Ci = 3.7x1010 BQ.

Q.42. Define Isomeric transition.

Ans. A mode of radioactive decay where a nucleus goes from a higher to lower energy stage. The mass number and atomic number remain unchanged.

Q.43. Define Nucleon.

Ans. A constituent of the nucleus that is proton or neutron

Q.44. Define Nuclear reactor.

Ans. A device in which a fission reaction can be intiated, maintained and controlled. Its essential components are fissionable fuel, moderator, shielding control rods and coolant.

Q.45. The transformation of one element into another by a nuclear reaction is called..........................

Ans. Transmutation

Q.46. What is Cherenkov radiation?

Ans. Light emitted by particles that move through a medium in which speed of light is slower than speed of particle.

Q.47. What are the isotopes used for soil water movement studies?

Ans. Tritium (^{3}H) - for water movement, ground water recharge and pollution studies. Deutrium (^{2}H) and oxygen (^{18}O) - Hydrological and Seepage studies.

Q.48. Name few elementary particles.

Ans. Protons, electrons, neutrons

Q.49. For an element. If atomic number - Z; Number of Neutrons = N then. What is the mass number (A)?

Ans. $A = Z + N$

Q.50. The amount of available nutrient measured in terms of fertilizer standard when plan root is exposed to two separate sources of nutrients is called ....

Ans. A value

Q.51. After 3 half-lives, the remaining activity of a radioisotope is .... of its original activity?

Ans. 1 / 8

Q.52. 'A' value technique was given by ....

Ans. Fried and Dean (1954)

Q.53. After two hours, 1/16th of the initial amount of radioactive isotope remains unchanged, the half life of the isotope is

Ans. 30 minutes.

Q.54. 'L' value was proposed by the scientist ....

Ans. Larsen (1952)

Q.55. 'E' value concept was given by the scientist ....

Ans. Russel et al. (1954)

Q.56. The nutrient availability concept that does not include the plant system is ....

Ans. E value.

Q.57. Name the rays emitted by radioactive substances.

Ans. α, β and γ rays

Q.58. Which rays have greatest penetrating power?

Ans. γ rays

Q.59. Write the charge of alpha rays.

Ans. Positively charged

Q.60. The half life of radium is 1600 years. After how much time it will remain 25% of the present?

Ans. 3200 years

Q.61. Mention few alpha emitters

Ans. Rubidium, Polonium, Uranium

Q.62. Name a radioactive inert gas.

Ans. Radon

Q.63. What is the number of electrons which each sub shell can occupy?

Ans. $2n^2$

Q.64. Which experiment shows the existence of the particles in the nucleus of an atom?

Ans. Alpha ray scattering experiment

Q.65. What would be the result of transition of an electron from M level to K level?

Ans. Emisson Spectra

Q.66. Write the electronic configuration of a noble gas.

Ans. 2, 8, 8

Q.67. Negative ions are formed from neutral atoms by gain of ....

Ans. Electrons

Q.68. An oxygen atom contains .... and ....

Ans. 8 protons and 8 electrons

Q.69. An electron is identical with........................particle

Ans. β - ray

Q.70. The emisson of β rays produces an

Ans. Isobar

Q.71. Atomic number and Atomic mass of uranium is 92 and 238. During the radio active disintegration it loses 5a and 4B particles. Find out the atomic number and weight of new element.

Ans. Atomic number = 86. Atomic weight 218

Q.72. $^{17}Cl_{35}$ and $^{17}Cl_{37}$ are ....

Ans. Isotopes

Q.73. Name a device used for the measurement of radioactivity

Ans. Geiger Muller Counter

Q.74. What is an alpha particle?

Ans. Alpha particles consist of two units of positive charges and having a mass of 4 units. Have little penetrating power.

Q.75. What are Beta particles?

Ans. Beta particles are high speed electrons emitted by unstable nuclei of the radioactive atoms at the time of its disintegration.

Q.76. Define gamma radiation.

Ans. Gamma rays are uncharged particles and not reflected in magnetic and electric fields. Have more penetration power than α and β particles.

❑❑❑

Chapter-9

REMOTE SENSING

Q.1. What is remote sensing?

Ans. Sensing of an object or a phenomenon without being in actual contact.

Q.2. What are the essential components of remote sensing?

Ans. Signal, Sensor, Sensing

Q.3. What is the remote sensing system that uses natural light?

Ans. Passive remote sensing

Q.4. What is active remote sensing?

Ans. A remote sensing system that uses own source of light. Eg. RADAR

Q.5. Who proposed the laws of electro dynamics?

Ans. James clerk Maxwell

Q.6. What are the theories governing EMR?

Ans. Wave theory and Particle theory

Q.7. What are the two types of scattering?

Ans. Elastic Scattering and Inelastic scattering

Q.8. What are the types of elastic scattering?

Ans. Raleigh scattering, Mie scattering and Non selective scattering

Q.9. What are the types of inelastic scattering?

Ans. Compton scattering and Raman scattering

Q.10. What causes the blue colour of the sky?

Ans. Rayleigh scattering

Q.11. What is the name of electromagnetic signals useful in satellite remote sensing which are least attenuated by atmospheric constituents?

Ans. Atmospheric windows.

Q.12. What are the types of Radar systems?

Ans. Imaging and Non-imaging

Q.13. Expand RADAR.

Ans. Radio Detection and Ranging

Q.14. Give an example for non imaging radar system.

Ans. Microwave scatterometer

Q.15. Give an example for Imaging radar system.

Ans. Side looking Airborne Radar (SLAR)

Q.16. What are the categories of SLAR.

Ans. Real Aperture Radar (RAR) and Synthetic Aperture Radar (SAR)

Q.17. Which remote sensing is unaffected by weather conditions and possible in both day and nights?

Ans. Radar or microwave remote sensing

Q.18. What are the two types of electromagnetic signal detectors?

Ans. Optical film detectors and optoelectronic detectors

Q.19. What is the sensor material in film detector?

Ans. Silver bromide

Q.20. What are the types of opto electronic detectors?

Ans. Thermal detectors and quantum detectors.

Q.21. Give an example for thermal detector.

Ans. Thermister bolometer

Q.22. What are the types of quantum detectors?

Ans. Photo emissive detectors, photo - conductive detectors, photo voltaic detectors.

Q.23. What are the types of sensor systems?

Ans. Non imaging sensor system and imaging sensor system.

Q.24. Give an example for non imaging sensor system.

Ans. Spectrometer

Q.25. What are the two types of Imaging sensor systems?

Ans. Framing system and scanning system.

Q.26. Give examples for framing system of imaging.

Ans. Photographic films and Return Beam Vidicon.

Q.27. Give examples for scanning systems of imaging.

Ans. Across track and along track scanning.

Q.28. What is the name of Across track scanner?

Ans. Whisk broom scanner

Q.29. What is Push broom scanner?

Ans. Along track scanner

Q.30. Which satellite uses across tract multispectral scanning?

Ans. Landsat series

Q.31. What is the image scale required for district level planning?

Ans. 1 : 50,000

Q.32. What are the types of resolution?

Ans. Spatial, Spectral, Temporal and Radiometric

Q.33. How is spatial information of an image obtained?

Ans. By use of imager, altimeters and sounders.

Q.34. How is spectral information of an image obtained?

Ans. By spectrometer

Q.35. How is the radio metric or intensity information of an image obtained?

Ans. Radiometers and Scatterometers

Q.36. What is the minimum distance between two objects that a sensor can record distinctly?

Ans. Spatial resolution

Q.37. What is minimum bandwidth which a sensor can resolve?

Ans. Spectral resolution

Q.38. What is radiometric resolution?

Ans. It is the number of discrete levels into which a signal may be divided (quantization).

Q.39. What is the time interval between two successive visits of a particular scene by remote sensor satellite?

Ans. Temporal resolution

Q.40. What are the types of remote sensor platforms?

Ans. Ground observation platform, airborne observation platform and space borne observation platform.

Q.41. Give examples for ground observation platforms.

Ans. Handheld photographic cameras and spectro radiometers, cherry pickers and portable masts.

Q.42. Give examples for air borne observation platforms.

Ans. Balloons, Aircraft and rockets.

Q.43. Expand NHAF Program.

Ans. National High Altitude Photography program (1978) coordinated by United States Geological Survey.

Q.44. Give an example for high altitude bounding rocket.

Ans. Skylark earth resource rocket

Q.45. What are space borne observation platforms?

Ans. Satellite platforms

Q.46. What are the types of satellite platforms?

Ans. Manned satellite platform and unmanned satellite platform.

Q.47. Give examples for manned satellite platforms.

Ans. Gemini, Apollo, Skylab, Space shuttle, International space station

Q.48. Give examples for unmanned satellite platforms.

Ans. Landsat series, SPOT series, IRS

Q.49. Give example for geo stationary satellite.

Ans. INSAT Series

Q.50. When did the space shuttle Columbia disaster take place?

Ans. February 2003.

Q.51. What are the types of Indian satellite launch vehicles?

Ans. PSLV and GSLV

Q.52. How are the orbits classified from the point of view of periodicity of satellite movement?

Ans. Geostationary and Sun synchronous

Q.53. What is the purpose of geostationary satellite communi-cation?

Ans. Television broadcast / Meteorological observations.

Q.54. Give examples of sun synchronous satellites.

Ans. LANDSAT series, SPOT series and IRS series.

Q.55. When was IRS - 1A launched?

Ans. 17th March 1988

Q.56. Which satellite was launched on 29th August 1991?

Ans. IRS - IB.

Q.57. What is the altitude and ground resolution of IRS - 1A. and IRS-1B?

Ans. 904 km, 72.5 m

Q.58. What is the altitude of IRS - 1C?

Ans. 817 km

Q.59. What is the area of the earth observed by satellite in one repeat cycle?

Ans. Coverage.

Q.60. What are the elements of coverage?

Ans. Nadir trace or ground track of satellite, sensor view area or swath

Q.61. What are the types of signal used in remote sensing?

Ans. Acoustic signal, Particulate signal, electromagnetic signal.

Q.62. Name the wavelengths normally used in remote sensing.

Ans. Visible and near infrared.

Q.63. What is the band used in near infra red?

Ans. 0.7-1.3 nm

Q.64. How will you calculate ratio of vegetation index?

Ans. Red / Near infrared

Q.65. Which spectral rays are used for mapping of radio active material?

Ans. Gamma rays

Q.66. What is the spectral rays used for vegetation properties?

Ans. Visible

Q.67. Define Band.

Ans. A range of wavelength of electromagnetic radiation is called band.

Q.68. What do you mean by Ground truth?

Ans. Database collected by field studies used to verify the correctness of remote sensing is known as Ground truth.

Q.69. Define kriging.

Ans. Kriging is an interpolation technique for obtaining statistically unbiassed estimates of surface elevations from a set of control points.

Q.70. Write the effects in reflectance for parameters used in remote sensing.

Ans. Parameter Effect in reflectance

Increase in

a) Soil moisture
b) Clay
c) Igneous rocks Decrease in reflectance
d) Organic matter
e) Iron oxide

Increase in

a) Sand content
b) Lime stone Increase in reflectance
c) Salinity

Increase in

Soil colour No effect in reflectance

Q.71. What are the format used to store the image data acquired from remote sensing system?

Ans. Band sequential, band interleaved by line.

Q.72. Define NADIR.

Ans. It is the point on the ground vertically beneath the respective centre of the camera lens.

Q.73. Define Pixel.

Ans. The smallest discrete element which make up a raster image is called as pixel.

Q.74. Define query.

Ans. Query is a method to find specific information in a data base

Q.75. What is Raster?

Ans. Raster is the scanned area of cathode ray tube.

Q.76. What do you mean by the term scanning?

Ans. Scanning is a process used to convert analog information from maps, photographs or overlays into a digital format by a computer.

Q.77. Expand GIS and GPS.

Ans. GIS - Geographic Information System

GPS - Global Positioning System

Q.78. What is thematic map?

Ans. Map related to a single topic theme or subject of discourse is called as thematic map.

Q.79. What are the functions of GIS in Cartography?

Ans. Data acquisition, mapping and display

Q.80. What are the uses of vector data structure and raster data?

Ans. *Vector data structure* : used for describing topology, simulation modeling.

Raster data structure : used for describing update, coordinate transformation

Q.81. Write the function of Digitizer.

Ans. Used for analog map kind of data source.

Q.82. Write the use of Stereo plotter.

Ans. Used for analog aerial photograph kind of data source.

Q.83. Name the equipment used for analog satellite imagery kind of data source.

Ans. Image zoom scope

Q.84. What is the function of GPS?

Ans. Used for analog ground survey kind of data source.

Q.85. Write the resolution of various scanners.

Ans. Very high resolution : Mechanical Scanner

Medium resolution : CCD Camera

Low resolution : Video Camera

Q.86. Define Spatial resolution.

Ans. It is a type of resolution that describes the capability of identifying nearly kept two objects.

Q.87. Define radiometric resolution.

Ans. It is a type of resolution that describes the sensitivity of the sensor to incoming radiance.

Q.88. Define the temporal resolution.

Ans. It is a type of resolution that describes the acquisition data recorded on different dates.

Q.89. What is spectral range.

Ans. The minimum distance between two objects that can be distinguished by continuous different wavelengh range is referred to as spectral range.

Q.90. What is spectral Signature?

Ans. The characteristic reflectance of an object.

□□□

Chapter-10

ANALYTICAL CHEMISTRY AND INSTRUMENTATION

Q.1. .... is the amount of matter contained in a substance and independent of gravitational force

a) Weight b) Mass
c) Density d) Volume

Q.2. .... is the force acting on mass and dependent on gravity

a) Weight b) Density
c) Mass d) Volume

Q.3. .... is the smallest particle of an element that takes part in chemical reaction

a) Molecule b) Position
c) Atom d) Neutron

Q.4. The number of protons in an atom is called as ....

a) Atomic number (Z) b) Atomic weight
c) Molecular weight d) Density

Q.5. The weight of 6.024×10^{23} numbers of atoms of any element is it's ....

a) Atomic weight b) Molecular weight
c) Molecular mass d) Density

Q.6. What is Avogadro's number?

a) $3.6x10^6$ b) 5 million

c) $6.024x10^{23}$ d) 10 million

Q.7. .... is the smallest particle of substance which retain the property of that substance and exist in free side

a) Atom b) Electron

c) Neutron d) Molecule

Q.8. Weight of one molecule of the substance relating to the weight of carbon atom taken as 12 is ....

a) Molecular weight b) Atomic weight

c) Equivalent weight d) Normal weight

Q.9. The weight of an element that combines or displaces 1.008 parts by weight of hydrogen or 8.00 parts by weight of oxygen or 35.5 parts by weight of chlorine is ....

a) Atomic weight b) Molecular weight

c) Normality d) Equivalent weight

Q.10. Equivalent weight of a substance expressed in grams is known as ....

a) Mole b) Gram equivalent weight

c) ppm d) Nulli equivalent/ litre

Q.11. One gram of equivalent weight of a solute dissolved in one litre of the solvent is one .... solution.

a) Molal b) Molar

c) Normal d) Formal

Q.12. One gram molecular weight of a solute dissolved in one litre of the solvent is one.................... solution

a) Molal b) Molar

c) Normal d) Formal

Q.13. One gram molecular weight of a solute dissolved in one kilogram of the solvent is one....................solution

a) Molal b) Molar
c) Normal d) Formal

Q.14. One tenth of (0.1) gram equivalent of a substance dissolved in one litre of the solvent is............ solution

a) Normal b) Molar
c) Decinormal d) Centinormal

Q.15. One hundredth (0.01) gram equivalent of a substance dissolved in one litre of the solvent is ....solution

a) Normal b) Molar
c) Decinormal d) Centinormal

Q.16. One gram of P dissolved in one litre of distilled water is..................solution

a) 100ppm b) 1000ppm
c) 1N d) 0.1N

Q.17. Number of gram equivalents of a solute dissolved in a number of litres of the solvent is known as ....

a) Normality b) Molarity
c) Molalitly d) ppm

Q.18. A 0.1 N solution of KOH is prepared by dissolving .... grams of KOH in 3 litre of distilled water

a) 5.6g b) 56g
c) 16.8g d) 168g

Q.19. A stock solution of one litre of 1000ppm K is available for preparing 100ml of 100ppm K. What is the volume of stock solution to be pipetted out?

a) 1 litre b) 100ml
c) 10ml d) 1ml

Q.20. A 20% solution (by weight) of KI is prepared by dissolving 20g of KI in................of water.

a) 80ml b) 80g
c) 80mg d) 80litre

Q.21. A 10% solution of ethanol (by volume) is prepared by mixing 10ml of ethanol with ....of water.

a) 90ml b) 90g
c) 90litre d) 90mg

Q.22. A 5% solution of NaoH(by strength) is prepared by dissolving 5g of NaoH in..........................of water

a) 95ml b) 95g
c) 100ml d) 1litre

Q.23. In a flame test, the following color shows the presence of copper

a) Lilac b) Brick red
c) Yellow d) Blue

Q.24. In a flame test the following color shows the presence of calcium

a) Lilac b) Brick red
c) Yellow d) Blue

Q.25. Lilac color in a flame test shows the presence of ....

a) Potassium b) Calcium
c) Sodium d) Lithium

Q.26. Yellow color in a flame test shows the presence of ....

a) Potassium b) Calcium
c) Sodium d) Lithium

Q.27. One of the following is a qualitative analysis

a) Volumetry b) Gravimetry
c) Flame test d) Potentiometry

Q.28. Luminescence analysis is used to determine .... of the substance.

a) Stability b) Flammability
c) Purity d) Ion selectivity

Q.29. ....is a method of quantitative analysis

a) Spot test
b) Luminescence
c) Micro crystallography
d) Gravimetry

Q.47. .... is a technique used to bring the constituents into solution, of the substances that are not soluble in a solvent

a) Distillation
b) Digestion
c) Titration
d) Standardization

Q.30. The most common method of separation and purification of liquid is........................

a) Digestion
b) Distillation
c) Ashing
d) Fusion

Q.31. ....method of sample preparation is used for large size samples.

a) Digestion
b) Distillation
c) Ashing
d) Fusion

Q.32. ....is a sample preparation technique for the sample analysis.

a) Digestion
b) Distillation
c) Ashing
d) Fusion

Q.33. One gram of formula weight of the solute dissolved in one litre of the solvent is one ....solution

a) Molar
b) Molal
c) Normal
d) Formal

Q.34. Triple acid mixture consists of Nitric, Sulphuric, Perchloric acids in the ratio of................

a) 9:2:1
b) 9:1:2
c) 1:9:2
d) 1:2:9

Q.35. Diacid mixture for N estimation consists of Sulphuric, Perchloric acids in the ratio of................

a) 9:2
b) 5:2
c) 7:2
d) 3:2

Q.36. Diacid mixture for S estimation consists of Nitric, Perchloric acids in the ratio of.......................

a) 9:2
b) 7:2
c) 5:2
d) 3:2

Q.37. ....ashing prevents volatilization of sulphur from sample.

a) Dry ashing
b) Wet ashing
c) Sulphated ashing
d) Combustion

Q.38. .... distillation is used to separate volatile organic compounds which have a vapour pressure of 1000 cdm or below and which are immiscible with water.

a) Steam distillation
b) Fractional distillation
c) Azeotopic distillation
d) Vaccum distillation

Q.39. .... distillation is used to separate liquids which as miscible with one another and with different boiling points

a) Steam distillation
b) Fractional distillation
c) Azeotopic distillation
d) Vaccum distillation

Q.40. Mixture of Benzene and water is separated by .... distillation

a) Steam
b) Fractional
c) Azeotropic
d) Vaccum

Q.41. ....distillation is used for liquids with very high boiling point

a) Steam
b) Fractional
c) Azeotopic
d) Vaccum

Q.42. One of the following is not a property of primary standard

a) Less purity
b) High solubility
c) Long shelf life
d) High equivalent weight

Q.43. For preparation of standard solutions,....................grade chemicals should be used

a) LR
b) CG
c) AR/GR
d) GR/LR

Q.44. The process of determining the strength of a titrate using standard solution is called....................

a) Distillation
b) Titration
c) Standardization
d) Combination

Q.45. ....is an example of back titration

a) Acidimetry
b) Alkalimetry
c) Complexometry
d) Thio cyanometry

Q.46. ....is an example of displacement titration

a) Iodometry
b) Iodimetry
c) Complexometry
d) Thio cyanometry

Q.47. In ....type of titrimetry, the titrate itself acts as an indicator

a) Iodometry
b) Dichrometry
c) Permanganimetry
d) Iodimetry

Q.48. .... is a substance which indicates the completion of a reaction

a) Buffer
b) Indicator
c) Titrant
d) Stabilizer

Q.49. A solution which resists sudden changes in pH is.............

a) Distilled water
b) Indicator
c) Alcohol
d) Buffer

Q.50. Ionic theory of indicators was proposed by................

a) Kjeldahl
b) Liebig
c) Ostwald
d) Mohr

Q.51. Nitrogen estimation method was proposed by............

a) Volhard
b) Mohr
c) Kjeldahl
d) Liebig

Q.52. Conversion of metal into metal ion complex is called ....

a) Neutralization
b) Precipitation
c) Reduction
d) Complexation

Q.53. ....is the process of gain of electrons by an atom or ion

a) Neutralization b) Sublimation
c) Hydration d) Reduction

Q.54. ....is the process of loss of electrons by an atom or ion

a) Oxidation b) Reduction
c) Hydration d) Complexation

Q.55. .... is the process of conversion of solid to vapour

a) Oxidation b) Reduction
c) Sublimation d) Hydrolysis

Q.56. ....is a light sensitive solution

a) Sodium chloride b) Silver nitrate
c) Potassium dichromate d) Sodium hydroxide

Q.57. ....is a strong oxidizing agent

a) Sodium chloride b) Potassium permanganate
c) Oxalic acid d) Sodium hydroxide

Q.58. Cuprimetric analysis is used in the estimation of ....

a) Barium sulphate b) Calcium
c) Reducing sugar d) Aluminium hydroxide

Q.59. Versenate titration or complexometry is used in the estimation of...............

a) Barium sulphate b) Calcium
c) Reducing sugar d) Aluminium hydroxide

Q.60. ....is a reducing agent

a) Oxalic acid b) Potassium permanganate
c) Potassium dichromate d) Sodium chloride

Q.61. In ....type of titrimetry, strength of KOH is determined with primary standard of succinic acid

a) Acidimetry b) Alkalimetry
c) Complexometry d) Argentometry

Q.62. In .... type of titrimetry, strength of HCl is determined with primary standard of sodium carbonate

a) Acidimetry b) Alkalimetry
c) Complexometry d) Argentometry

Q.63. .... type of titrimetry involves silver compounds

a) Acidimetry b) Alkalimetry
c) Complexometry d) Argentometry

Q.64. Estimation of halide is done by ....

a) Acidimetry b) Alkalimetry
c) Complexometry d) Argentometry

Q.65. The titration which involves transfer of electrons is ....

a) Acidimetry b) Alkalimetry
c) Dichrometry d) Argentometry

Q.66. One of the following is a precipitometric titration

a) Acidimetry b) Argentometry
c) Alkalimetry d) Complexometry

Q.67. An example for self indicator is....................

a) $KMnO_4$ b) $K_2Cr_2O_7$
c) KOH d) $AgNO_3$

Q.68. The indicator which is directly added to the solution being titrated is called as.......................

a) Self indicator b) Internal indicator
c) External indicator d) Universal indicator

Q.69. An example for internal indicator...............

a) $KMnO_4$ b) Phenolphthalein
c) Potassium ferri cyanide d) Potassium dichromate

Q.70. An example for external indicator................

a) $KMnO_4$ b) Methyl orange
c) Potassium ferri cyanide d) Potassium dichromate

Q.71. An example for indicator which forms colored complex with metal ion is................

a) Murexide b) Methyl orange
c) Starch d) $KMnO_4$

Q.72. The following theory of indicator states that color change of indicators is due to intermolecular rearrangement and change in the structure of indicators

a) Ostwold b) Liebig
c) Chromophore d) Ionic

Q.73. The indicator used in Dichrometric titration is ....

a) Potassium chromate b) Methyl orange
c) Phenolphthalein d) Diphenylamine

Q.74. An indicator employed in qualitative analysis is ....

a) Litmus b) Methyl orange
c) Diphenylamine d) Starch

Q.75. Eriochrome Black T is an indicator used in........................

a) Complexometry b) Iodometry
c) Dichrometry d) Alkalimetry

Q.76. The process of determining volume of a reagent required to complete the reaction with a known volume of an analyte is called ....

a) Digestion b) Titration
c) Standardization d) Distillation

Q.77. The most suitable indicator for the titrations involving weak acid Vs strong base is..................

a) Phenolphthalein b) Methyl orange
c) Murexide d) Methyl blue

Q.78. ....is the most commonly used metal chelating agent

a) EDTA b) Succinic acid
c) Sodium chloride d) Silver nitrate

Q.79. Solution of accurately known strength used in titration are called..........

a) Analyte b) Titrant
c) Indicator d) Buffer

Q.80. Weight of an analyte can be determined by.................

a) Potentiometer b) Coulometry
c) Gravimetry d) Voltametry

Q.81. The process of reversible and inter conversion of isometric forms of indicator is called as...........

a) Tautomerism b) Chromo morphism
c) Auxochromism d) Dimorphism

Q.82. ....is the point of completion of reaction when the solution is titrated with another solution

a) Equivalence point b) End point
c) Standardization d) Titration

Q.83. .... is the point of completion of reaction observed by the visual color change when the solution is titrated with another solution

a) Equivalentice point b) End point
c) Standardization d) Titration

Q.84. Gravimetry is based on ....measurement

a) Weight b) Volume
c) Density d) Height

Q.85. For a precipitate to form, the value of ionic product should exceed the..............

a) Solubility product b) Complexation effect
c) Ionic coefficient d) Dilution factor

Q.86. The solubility of silver chloride precipitate is decreased due to adding of KCl as precipitant is because of....................

a) Common ion effect b) Diverse ion effect
c) Dilution effect d) Ionic equilibrium

Q.87. Ripening of precipitate leads to ....

a) Aging
b) Transformation into stable forms
c) Formation of larger crystals
d) All the above

Q.88. Colored constituent of an analyte determined by relative absorption of light is the principle of ....

a) Potentiometry
b) Conductometry
c) Colorimetry
d) Coulometry

Q.89. Colorimetry is based on.................law

a) Ohm
b) Liebig
c) Beer Lambert
d) Faraday

Q.90. Potentiometry is based on .... equation

a) Ohm
b) Liebig
c) Lambert
d) Nernst

Q.91. Law of volumetric analysis is given as ....

a) $V_1N_1=V_2N_2$
b) V=Mass/Density
c) $E=mc^2$
d) F=ma

Q.92. The following is an optical method based on absorption in instrumental analysis

a) Voltametry
b) Conductometry
c) Coulometry
d) Colorimetry

Q.93. Measurement of charged gas molecules according to their mass is called............

a) Gas chromatography
b) Spectrophotometry
c) Flame photometry
d) Mass spectrometry

Q.94. .... law forms the basis for conductometry

a) Ohm
b) Liebig
c) Beer Lambert
d) Mohr

Q.95. The following is an electrical method in instrumental analysis..........

a) Coulometry
b) Colorimetry
c) Fluorimetry
d) Gravimetry

Q.96. The measurement of resistance offered by a substance is ....

a) Conductometry b) Potentiometry
c) Voltometry d) Coulometry

Q.97. .... instrument works on the basis of separation due to mobility

a) Gas liquid chromatography b) Potentiometer
c) Flame photometer d) Colorimeter

Q.98. High equivalent weight is desirable for substance used as a primary standard because ....

a) It is stable b) Weighing error minimized
c) It is not hygroscopic d) It contains less impurity

Q.99. The instrument which uses the principle of rotation of plane of polarization by an optically active principle is...........

a) Voltmeter b) Potentiometer
c) Polarimeter d) Flame photometer

Q.100. The isolation of desired analytical wavelength of radiation for passage through the detector is done by................

a) Nebuliser b) Flame assembly
c) Printer d) Monochromator

Q.101. Estimation of micronutrients and heavy metals is done by using .... instrument

a) Potentiometer b) Atomic absorption spectrophotometer
c) Polarimeter d) Flame photometer

Q.102. The visual output of chromatography is called............

a) Mammogram b) Monogram
c) Chromatogram d) Electrogram

Q.103. The characteristic time taken for a particular analyte to pass from the column inlet to the detector under specified condition in chromatography is termed as ....

a) Absorption time b) Passage time
c) Moving time d) Retention time

Q.104. An instrumental method based on scattering of radiation is ....

a) Conductometry
b) Nephelometry
c) Polarimetry
d) Coulometry

Q.105. One of the following is not a sorption mechanism of chromatography

a) Surface adsorption
b) Ion exchange
c) Partition
d) Emission

Q.106. An instrument used for measurement of radio activity of a given substance is ....

a) GM counter
b) Flame photometer
c) Colorimeter
d) Infrared spectrometer

Q.107. One of the following is a crude error

a) Uncalibrate pipette
b) Spilling of precipitate
c) Over weighing of precipitate
d) Coprecipitate

Q.108. Which is a random error?

a) Uncalibrate pipette
b) Spilling of precipitate
c) Coprecipitate
d) Temperature fluctuation

Q.109. Absolute error expressed as a fraction of original value is

a) Random error
b) Methodological error
c) Crude error
d) Relative error

Q.110. One of the following is not true about distillation

a) Uniform heating
b) Use of round bottom flask
c) Main fraction collected separately
d) Facilitate loss of vapour

Q.111. The errors which are beyond the control of analyst are ....

a) Personal error
b) Instrumental error
c) Random error
d) Methodological error

Q.112. Blank determination is a technique to minimize error due to ....

a) Reagent impurities
b) Instrumental error
c) Faulty method
d) Skill of analyst

Q.113. The nearness of analytical value to the true value is called ....

a) Error
b) Accuracy
c) Precision
d) Analysis

Q.114. The reproducibility of results of analysis under identical experimental condition is called ....

a) Error
b) Precision
c) Accuracy
d) Analysis

Q.115. Precision of analytical results depends on...............

a) Instrumental error
b) Methodological error
c) Personal error
d) Indeterminate error

Q.116. The volumetric analysis does not satisfy the following condition

a) Fast reaction
b) Stoicchio metric
c) Side reaction
d) Marked changed at end point

Q.117. Which of the following is another way to express a KCl solution 0.01% by weight

a) 1.0M
b) 10ppm
c) 100ppm
d) 1N

Q.118. ....is a separation method that relies on difference in partitioning behavior between a mobile phase and a stationary phase to separate the components in a mixture

a) Gravimetry
b) Volumetry
c) Colorimetry
d) Chromatography

Q.119. In a high performance liquid chromatography ....is a mobile phase

a) Solid
b) Semi solid
c) liquid
d) Gas

Q.120. Law of mass action was entitled by..............

a) Ostwald
b) Gulberg & Waage
c) Mohr
d) Lindsay & Norvell

Q.121. The "law of mass action" states that the rate of a chemical reaction is proportional to the................of the reacting substances at constant temperature

a) pH
b) Equivalent weight
c) Density
d) Active mass

Q.122. The product of total molecular concentration of the ions of a sparingly soluble salt with the coefficient raised as their respective power is a constant at constant temperature and it is called as ....

a) Activity coefficient
b) Dissociation constant
c) Equilibrium constant
d) Solubility product

Q.123. The reaction involving electron transfer is ....

a) Reaction between $FeSO_4$ & $K_2 Cr_2 O_7$
b) Reaction between HCl & NaOH
c) Reaction between EDTA & Ca CO_3
d) Reaction between succinic acid & KOH

Q.124. A................normal solution of KOH is prepared by dissolving 5.6g of KOH in 1 litre of distilled water

a) One
b) Deci
c) Centi
d) Milli

Q.125. The indicator selected for a titration exhibit a distinct colour changes at a pH to that obtaining at...............

a) Equivalence point
b) Acidic pH
c) Alkaline pH
d) Ionization point

Q.126. ....indicator change from predominantly acid to a predominantly alkaline colour within a small interval of pH

a) Neutralisation indicator
b) Redox indicator
c) Precipitation indicator
d) Complexometric

Q.127. A solution containing milligram equivalent (1/1000g equivalent) of a substance in a litre of solution is expressed as ....

a) ppm b) ppb
c) pph d) meq/litre

Q.128. The atomic weight of an element is expressed in ....

a) ppm b) kg
c) a.m.u d) grams

Q.129. ....hypothesis states that equal volume of all gases under S.T.P (standard temperature and pressure) contains equal number of molecules

a) Liebig b) Nernst
c) Faraday d) Avogadro

Q.130. The electrical conductivity of 0.01 N KCl at 250C is ....

a) 1.14 b) 1.41
c) 2.2 d) 3.1

Q.131. The standard equivalent factor or volumetric factor for 1ml of 1N H_2SO_4 is equal to ....

a) 14g N b) 1.4gN
c) 0.14 gN d) 0.014g

Q.132. The negative logarithm value for expressing the concentration of H+ or OH- ions was suggested by Sorenson because

a) It is cumbersome to express the small numerical value of H^+ or OH^-
b) Simple whole number can be used
c) pH can be expressed from 0-14 as fixed points.
d) All the above

Q.133. The most accurate method of determining the pH of a solution depends on .... measurements

a) Density b) Ionization
c) Electro motive force d) Gravitational force

Q.134. The conductance of a solution enclosed between two electrode of 1 sq cm area and 1 cm apart is called ....

a) Specific conductance
b) Specific resistance
c) Equivalent conductance
d) Molar conductance

Q.135. The conductance of a solution containing 1g equivalent of dissolved electrolyte placed between two electrodes 1cm apart is termed as................

a) Specific conductance
b) Specific resistance
c) Equivalent resistance
d) Equivalent conductance

Q.136. The conductance of solution containing 1g molecules of dissolved electrolyte placed between two electrodes 1cm apart is termed as...........

a) Specific conductance
b) Specific resistance
c) Equivalent conductance
d) Molar conductance

Q.137. Conductivity cells are fitted with platinum electrodes to minimize the.....................effect

a) Polarization
b) Conductance
c) Refraction
d) Scattering

Q.138. The commonly used method of measuring resistance was invented by ....

a) Faraday
b) Liebig
c) Charles wheat stone
d) Edison

Q.139. One of the following criteria does not hold good for a satisfactory colorimetric analysis..........

a) Proportionality between color and concentration
b) Specificity of color reaction
c) Stability of color
d) Less sensitivity of colour reaction

Q.140. In spectrophotometer determination a standard or reference or calibration curve is drawn relating the concentration of standards to the .... values

a) Absorbance
b) Scattering
c) Emission
d) Ionization

Q.141. Elementary quantum theory was proposed by ....

a) Nernst b) Ohm
c) Faraday d) Max planck

Q.142. An example for a chemical change of a substance is

a) Rusting of iron b) Digestion of food
c) Burning of residues d) All the above

Q.143. An example for a physical change of a substance is

a) Hammering of metal b) Freezing or boiling of water
c) Both a & b d) Neither a nor b.

Q.144. Chemical change which involves break down of a compound into simpler compounds or elements is called as ....

a) Replacement b) Combination
c) Decomposition d) Deposition

Q.145. Chemical change which involves substitution of one element for another is called as ....

a) Replacement b) Combination
c) Decomposition d) Double decomposition

Q.146. ....gave the definition that acid is a substance that gives up a proton and a base is a substance that accepts a proton

a) Planck b) Faraday
c) Lowery Bronsted d) Lewis

Q.147. According to ....definition, a base is substance that gives up an electron pair to form a co-ordinate bond and an acid is a substance that can take up an electron pair to form the bond

a) Plank b) Nernst
c) Lewis d) Faraday

Q.148. An example of a salt of weak acid and a strong base is ....

a) Ammonium chloride b) Ammonium hydroxide
c) Sodium chloride d) Sodium acetate

Q.149. An example for salt of a strong acid and weak base is ....

a) Ammonium chloride
b) Sodium acetate
c) Sodium chloride
d) Ammonium acetate

Q.150. An example for salt of strong acid and strong base is ....

a) Ammonium chloride
b) Sodium acetate
c) Sodium chloride
d) Ammonium acetate

Q.151. An example for a salt of weak acid and weak base is ....

a) Ammonium chloride
b) Sodium acetate
c) Sodium chloride
d) Ammonium acetate

Q.152. ....is not a suitable indicator in potentiometry electrode due to hardness and brittleness

a) Silver
b) Copper
c) Lead
d) Tungsten

Q.153. .... is an example for absorption indicator used in argentometry

a) Dichloro fluorescein
b) Diphenylamine
c) Bromo erosol green
d) Phenolphthalein

Q.154. The following is the important deviation of Beer Lambert law

a) Does not hold good at high concentration of analyte
b) Does not hold good if nature of species change with dilution or pH
c) Does not hold good in system involved ionization equilibria
d) All the above

Q.155. .... device is used to determine the intensity of illumination in the infra red radiation

a) Thermister
b) Thermo couple
c) Bolometer
d) Conductometer

Q.156. A device used to determine changes in temperature is called ....

a) Conductometer
b) Potentiometer
c) Thermostat
d) Colorimeter

Q.157. The physical property considered during chromatographic separation of mixture is ....

a) Solubility
b) Adsorption
c) Volatility
d) All the above

Q.158. In chromatography if both the moving and stationary phase are liquid then it is termed as..............

a) Partition chromatography
b) Adsorption chromatography
c) Gas chromatography
d) Ion exchange chromatography

Q.159. In chromatography, if moving phase is a liquid and stationary phase is solid functioning as an adsorbing surface then it is called as ....

a) Partition chromatography
b) Adsorbtion chromatography
c) Gas chromatography
d) Ion exchange chromatography

Q.160. In a chromatography, if the moving phase is a gas and stationary phase is a liquid it is called as ....

a) Ion exchange chromatography
b) Thin layer chromatography
c) Gas chromatography
d) HPLC

Q.161. ....is the most extensively used TLC adsorbent

a) Silica gel
b) Benzene
c) Toluene
d) Ether

Q.162. Quantitative analysis is done in chromatography is done by....................method

a) Planimetry
b) Triangulation
c) Correlation of sample peaks with standard
d) All the above

Q.163. For determining the total soluble salts in water, the following parameter needs to be analysed

a) pH
b) EC
c) SAR
d) RSC

Q.164. Total Soluble Salts of irrigation water = EC x ....

a) 640 b) 10
c) 500 d) 1000

Q.165. Sulphate content in irrigation water is determined by........... method

a) Turbidimetry b) Chromatography
c) Flame photometry d) Potentiometry

Q.166. A diagram representing the alkali hazard of irrigation water was given by ....

a) Wilcox b) Liebig
c) White ny d) Mohr

Q.167. Crude protein content is obtained by multiplying the nitrogen content by a factor of ....

a) 1.25 b) 6.25
c) 0.0014 d) 10.25

Q.168. The total solids in cane juice is estimated by .... device

a) Brix hydrometer b) Colorimeter
c) Flurometer d) Spectro photometer

Q.169. A device which gives both sugar percentage and refractive index in a drop of clarified cane juice is............

a) Colorimeter b) Abbe refractometer
c) Saccharimeter d) Polarimeter

Q.170. Determination of ascorbic acid in vegetable is done by titration with ....

a) 2-6 di chlorophenol indophenol dye
b) Potassium permanganate
c) KOH
d) Oxalic acid

❑❑❑

ANNEXURES

Annexure I
Forms of Nutrient Elements Absorbed by Plants

Nutrient element	Forms absorbed
K	K^+
Ca	Ca^{2+}
Mg	Mg^{2+}
Fe	Fe^{2+} (Ferrous), Fe^{3+} (Ferric)
Mn	Mn^{2+} (Manganous), Mn^{3+} (Manganic)
Cu	Cu^{2+} (Cuprous) Cu^{3}+ (Cupric)
Zn	Zn^{2+}
Cl	Cl^-
CO	CO^{2+}
Na	Na^+
N	NH^{4+}, NO^{3-}
P	$H_2PO_4^-$, HPO_4^{2-}
S	SO_4^{2-}
B	H_3BO_4 (Boric acid) $H_3BO_3^-$ (Borate)
Mo	MoO_4^{2-} (Molybdate)
C	CO_2 (Mostly through leaves)
O	CO_2, O_2^{2-} OH^-, CO_3^{2-}
H	H^+, H_2O
Water	H^+, OH^-

Annexure II
Fertilizer Nutrient Contents

i) N Fertilizer

	Fertilizer	N content (%)
1.	Urea	46
2.	Mono ammonium phosphate	11-13
3.	Di ammonium phosphate	18
4.	NH_4C1	25
5.	Calcium Ammonium Nitrate (CAN)	25-26
6.	$NaNO_3$	16
7.	$Ca(NO_3)_2$	15.5
8.	Anhydrous NH_3	99
9.	$NH_4\ NO_3$	33.0
10.	$(NH_4)_2\ SO_4$	20.6
11.	KNO_3	13
12.	$(NH_4)_2\ SO_4 + NH_4\ NO_3$	26
13.	Calcium cyanamide	21
14.	Nitro Phosphate	23
15.	Ammonium Thiosulphate	12
16.	Urea Ammoniam Nitrate (Liquid)	32

ii) Slow release N Fertilizer

	Fertilizer	N content (%)
1.	Urea Formaldehyde	38
2.	Oxamide	32
3.	Thiourea	37
4.	IBDU	32
5.	Urea granular neem Coated	46

iii) P Fertilizers

	Fertilizer	P_2O_5 content (%)
1.	SSP	14-16
2.	TSP	46 - 48
3.	Mono ammonium phosphate	48 - 62
4.	Di ammonium phosphate	46
5.	Basic slag	14 -18
6.	Di calcium phosphate	34 - 39
7.	Rock phosphate	20 - 40
8.	Raw bone meal	20 - 25
9.	Steamed bone meal	22
10.	Nitro phosphate	11.5
11.	Super Phoshphoric acid (Liquid)	70

iv) K Fertilizers

	Fertilizer	K_2 Content (%)
1.	Muriate of potash	60
2.	K_2SO_4 Sulphate of Potash	50
3.	Potassium Schoenite	23

v) Ca Fertilizers

	Fertilizer	Ca (%)
1.	SSP	18 - 21
2.	TSP	12 - 14
3.	CAN	8.0
4.	Ca $(NO_3)_2$	19
5.	Ca - EDTA	3 - 5
6.	Gypsum ($CaSO_4$ - $2H_2O$)	30
7.	Basic slag	20

vi) Mg Fertilizers

	Fertilizer	Mg (%)
1.	Epsom salt ($MgSO_47H_2O$)	16
2.	Kiescrito ($MgSO_4H_2O$)	27
3.	Magnesium Nitrate	16
4.	Magnesium Chloride	8 - 9
5.	Mg Chelate	2 - 9
6.	Magnesium Sulphate	9.6
7.	Basic slag	3 - 4
8.	Mg-EDTA	2 - 4
9.	Magnesium Chloride ($MgCl_2.10H_2O$)	8 - 9

vii) S Fertilizers

	Fertilizer	S (%)
1.	Elemental Sulphur	90
2.	Gypsum	12 - 18
3.	$NH_4(SO_4)_2$	21
4.	TSP	1-1.5
5.	SSP	11
6.	Sulphur Coated urea	38
7.	Ammonium thio sulphate	26
8.	Zinc Sulphate	11
9.	Manganese sulphate	19.5
10.	Copper sulphate	13
11.	Ferrous sulphate	13
12.	K_2SO_4	18.5

viii) Micronutrient Fertilizers

	Fertilizer	Micronutrient content (%)
1.	Zinc sulphate	21.0% Zn
2.	Manganese sulphate	30.5% Mn
3.	Copper sulphate	24.0% Cu
4.	Ferrous sulphate	19.0% Fe
5.	Ammonium Molybdate	52.0% Mo
6.	Borax (Sodium tetra borate)	10.5% B
7.	Chelated Zn (Zn-EDTA)	12% Zn
8.	Chelated Fe (Fe-EDTA)	12.5% Fe
9.	Zn Sulphate Monohydrate	33% Zn
10.	Boric acid	17% B

ix) Complex Fertilizers

	Fertilizer	Nutrient Content
1.	DAP	18%N, 46% P_2O_5
2.	Ammonium Phosphate Sulphate (16:20:0:13)	16%N, 20% P_2O_5 13% S
3.	Ammonium Phosphate Sulphate (20:20:0:13)	20%N, 20% P_2O_5 13% S
4.	Nitro Phosphate (20:20:0)	20%N, 20% P_2O_5
5.	Mono Ammonium Phosphate	11%N, 52% P_2O_5

x) Fortified fertilizer

	Fertilizer	Nutrient Content
1.	Boronated Single Super Phosphate	16% P_2O_5 0.15 - 0.20 % B
2.	Zincated Urea	43%N, 2% Zn
3.	DAP Fortified with Boron	18%N, 46% P, 13% B
4.	Calcium Nitrate with Boron	14.6%N, 17.1%, Ca 0.25% B

xi) 100% Water Soluble Complex fertilizer

	Fertilizer	Nutrient Content
1.	Potassium Nitrate	13%N, 45% K_2O
2.	Mono Potassium Phosphate	52% P_2O_5, 34% K_2O
3.	Calcium Nitrate	15.5%N, 18.8% Ca
4.	Mono Ammonium Phosphate	12%N, 61% P_2O_5

As per the FCO Specification Nutrient Contents in Organic Manures

	Fertilizer	Nutrient Content
1.	City Compost	12% Organic Carbon 0.8% N 0.4% P_2O_5, 0.4% K_2O C:N ratio < 20
2.	Vermi compost	18% Organic Carbon

Annexure III
Nutrient Content in Plants

Nutrient element	Concentration
N	1.0 - 3.0%
P	0.2 %
K	1.0 %
Ca	1.0 - 4.0%
Mg	0.05 - 1.0%
S	0.1%
Fe	10 - 1000 ppm
Mn	50 ppm
Zn	5 -100 ppm
Cu	20 ppm
Mo	0.01 - 10 ppm
Cl	0.1 ppm
B	2-70 ppm

Annexure IV
Formula, Calculations, Conversions

Q.1. What is the factor to convert P % to P_2O_5 and vice versa.

Ans. % P = % P_2O_5 x 0.43

% P_2O_5 = %P x 2.29

Q.2. What is the factor to convert % K to % K_2O and vice versa.

Ans. % K = % K_2O x 0.83

%K_2O= % K x 1.2

Q.3. State stoke's Law.

Ans. *Stoke's Law* : Velocity of fall of a particle with the same density is directly proportional to the square of the radius and inversely proportional to the viscosity of the medium.

$$V = \frac{2}{9}\frac{(dp-d)gr^2}{\eta}$$

V = Velocity of fall (cm / sec)
G = acceleration due to gravity (cm / sec²)

$$dQ = DA\left(\frac{dc}{dx}\right)dt$$

dp = density of the particle (g / cc)
d = density of the liquid (g / cc)
r = radius of the particle (cm)
h = absolute viscosity of the liquid (poise)

Q.4. Write the formula to calculate bulk density.

Ans. $\text{Bulk density} = \frac{\text{Wt. of soil mass}}{\text{Soil volume}} \text{g / cc}$

Q.5. Write the formula to calculate porosity (or) % pore space.

Ans. $\left[1 - \frac{\text{Bulk density}}{\text{Particle density}} \times 100\right]$

Q.6. State Fick's law

Ans. *Fick's law* : Diffusion is a function of the concentration gradient, diffusion co-efficient of the medium and the cross sectional area.

dQ = mass flow (moles) during time t

Q.7. What is the pH of a solution having H+ = 4.2 x 10^{-6} g ions litre^{-1}?

Ans. pH = -logH+
= -log (4.2x10^{-6})
= - 0.62 - (-6)

pH = - 0.62 + 6 = 5.32

Q.8. What is H+ ion concentration if a Solution has pH= 4.38?

Ans. H+ = 10pH
= 10-4.38
= 100.62x10^{-5}
= 4.2x 10^{-5} g ions litre^{-1}

Q.9. Define Neutralizing value (or) Calcium Carbonate Equivalent (CCE).

Ans. CCE is defined as the acid neutralizing capacity of liming material expressed as a weight percentage of $CaCO_3$.

$$\text{CCE of a liming material} = \frac{\text{Molecular wt of } CaCO_3}{\text{Molecular wt. of liming material}} \times 100$$

whose CCE is to be determined

Q.10. Calculate the CCE of dolomite.

Ans. $$\text{CCE of dolomite} = \frac{\text{Molecular wt of } CaCO_3}{\text{Molecular wt. of dolomite}} \times 100$$

$$\text{Mole. Wt. dolomite} = \frac{184}{2} = 92$$

$$CaMg(CO_3)\text{CCE of dolomite} = \frac{100}{92} \times 100 = 108.7$$

Q.11. Write the formula for per cent effective $CaCO_3$ (ECC) (or) Neutralizing index.

Ans. Percent ECC (or) NI = CCE x fineness factor

Q.12. Write the formula for Base saturation percent.

Ans. $$\text{Percent base Saturation (BS)} = \frac{\text{me of basic cations}}{\text{Total Exchangeable cations}} \times 100$$

Q.13. Write the formula for calculating osmotic pressure of soil solution.

Ans. Osmotic pressure = 0.36 x EC
(in Atm (or) bar)

Q.14. Write the formula to calculate Sodium Adsorption Ratio (SAR).

Ans.

$$SAR = \frac{Na^+}{\sqrt{\frac{Ca^{2+} + Mg^{2+}}{2}}}$$

Na^+, Ca^{2+}, Mg^{2+} are concentrations of respective ions in saturation extract in mel^{-1}

Q.15. Write the formula to calculate Exchangeable Sodium Per centage (ESP).

Ans.

$$ESP = \frac{100\ (-0.0126 + 0.01475\ SAR)}{1 + (-0.0126 + 0.01475\ SAR)}$$

Q.16. Write the formula to calculate Leaching Requirement (LR).

Ans.

$$LR = \frac{Ddw \times 100}{Diw} = \frac{ECiw \times 100}{ECdw}$$

Ddw = Depth of drainage water in inches
Diw = Depth of irrigation water in inches
EC iw = EC of irrigation water (dsm^{-1})
EC dw = EC of drainage water (dsm^{-1})

Q.17. Give the formula to calculate Gypsum Requirement (GR).

Ans.

$$\text{Gypsum requirement (GR)} = \frac{\text{ESP (initial) -ESP (Final)}}{100} \times CEC$$

(me of Ca^{2+} / 100 g of soil)

Q.18. Write the formula to calculate salt index.

Ans. Salt index = (Total Na-24.5)-(Total Ca-Ca as Ca CO_3) x 4.85

Irrigation water of high quality have -24.5 to 0. Water having positive value of salt index is harmful for irrigation purposes.

Q.19. Write the formula to calculate Residual Sodium Carbonate (RSC).

Ans. RSC (me l^{-1}) = (CO_3^{2-} + HCO_3^-) (Ca^{2+} + Mg^{2+})

Q.20. Write the formula to calculate Soluble Sodium Percentage (SSP).

Ans. $$SSP = \frac{Na}{Ca + Mg + Na + K} \times 100$$

All soluble cations are expressed in me l^{-1}

Q.21. Write spillman's growth equation.

Ans. Y = M(l-Rx)

Y - Yield, M - Maximum yield, x = Growth factor

R = Constant

Q.22. Write Mitscherlich - Bray equation of growth.

Ans. $Log(A\text{-}Y) = \log A\text{-}C_1 b\text{-}C_x$

A - Maximum yield Y- yield obtained

X - amount of added nutrient

C1 - Efficiency factor of *b*

C - Efficiency factor of x

Q.23. Write the formula to calculate Agronomic efficiency.

Ans. $$\text{Agronomic efficiency} = \frac{\left|\begin{matrix}\text{Grain yield in}\\ \text{Fertilized plot}\end{matrix}\right| - \left|\begin{matrix}\text{Grain yield in}\\ \text{Control plot}\end{matrix}\right|}{\text{Amount of Nutrient applied}}$$

$$\text{Apparent N recovery} = \frac{\left|\begin{matrix}\text{Uptake of N in}\\ \text{fertilized plot}\end{matrix}\right| - \left|\begin{matrix}\text{Uptake of N in}\\ \text{Control plot}\end{matrix}\right|}{\text{Amount of Nutrient applied}} \times 100$$

$$\text{Physiological recovery} = \frac{\left|\begin{matrix}\text{Grain yield in}\\ \text{fertilized plot}\end{matrix}\right| - \left|\begin{matrix}\text{Grain yield in}\\ \text{Control plot}\end{matrix}\right|}{\left|\begin{matrix}\text{Uptake of nutrient}\\ \text{in fertilized plot}\end{matrix}\right| - \left|\begin{matrix}\text{Uptake of nutrient}\\ \text{in Control plot}\end{matrix}\right|}$$

Q.24. What are the units of radio activity.

Ans. 1 Curie = 3.7 x 10 rd

1 becquerel = 1 dps

Q.25. Give formula for 'A' vlaue.

Ans. 'A' value concept was developed by Fried and Dean (1954)

$$A = \frac{B(1-y)}{y}$$

A - Amount of available nutrient in soil

B - Amount of applied fertilizer nutrient

Y - Proportion of nutrient in the plant derived from the fertilizer nutrient

Q.26. Write formula for 'L' value.

Ans. 'L' value was suggested by Larsen (1952)

'L' value is used to calculate the P supply of the soil

$$\text{L value} = K\frac{x\,(Co - C)}{C}$$

L - available P in soil

x - Amount of P applied to soil

Co - K x. total P in plant

C - K x P uptake by plant from X

K - Proportionality constant

Q.27. State Darcy's law.

Ans. Rate of flow increased with an increased depth of water above the bottom of soil and decreased with an increased depth of soil through which water flowed.

$$Qw = -Kx\frac{(\Delta dw)\Delta t}{(\Delta ds)}$$

Qw - Quantity of water in cc

K - Hydraulic conductivity (cm / sec)

Δw - Water or hydraulic head (cm)

A - Soil area (Sq. Cm)

T - time in (sec)

ds - Soil depth (cm)

Q.28. Write De bye-huckel equation.

Ans. Theory of estimating activity coefficients of an ion is based on law of electrostatics and thermodynamics.

$$\text{Log } \delta \text{ i} = - \text{AZi}^2\text{W}^{1/2}$$

i - activity Coefficient
A - Constant varies with temperature
Zi - Valency of the ion
M - Ionic strength

Q.29. Give equation for diffusion coefficient.

Ans. Diffusion coefficient controls the movement of ions from soil to roots.

$$De = \text{Dw } \theta f / \text{ b}$$

De - Effective diffusion Coefficient
Dw - Diffusion Coefficient of ions in water
θ - Volumetric water percentage
f - Tortuosity factor
b - buffering capacity of soil

Q.30. State Freundlichs equation for soil phosphorus adsorption studies.

Ans. $\text{m} = \text{AC}^{\text{B}}$

m - amount of P ions adsorbed per unit weight of soil
C - Concentration of P ions in solution
A, B - Constants

Q.31. Write Gapon's equation.

Ans. *Monovalent* - Divalent cation exchange can be expressed as follows known as Gapon's equation

$$\frac{\text{C+ ads}}{\text{C}^{2+}\text{ads}} = \text{K}\frac{\text{a}^{+}\text{c}}{\sqrt{\text{a}^{2+}\text{c}}}$$

$$\frac{\text{a}^{+}\text{c}}{\sqrt{\text{a}^{2+}\text{c}}} = \text{AR}$$

C^{+} ads, C^{2+} ads - adsorbed monovalent, divalent cations
a^{+}c, a^{2+}c - activity of monovalent, divalent cations (mol l-1]
AR - Activity ratio
K - Gapon's Coefficient or selectivity Coefficient

Q.32. Write the formula to calculate heat capacity of a material.

Ans. Heat capacity C = S x m

S - Specific heat m - mass (calories]

Q.33. Write Hofmeister's cation sequence.

Ans. It is relative replacing power of particular cation species which depends on its strength of bonding

$Li^+ < Na^+ < K^+ < Rb^+ < Cs^+ < Mg^{2+} < Ca^{2+} < Sr^{2+} < Ba^{2+}$ increase in relative replacing power.

Q.34. Write the formula to calculate ionic strength.

Ans. $M = \frac{1}{2} CiZi^2$

Ci - Concentration of ions (i) in moles
Zi - Valency of ions

Q.35. Write Langmuir's adsorption equation.

Ans. $\frac{C}{x} = \frac{1}{Kb} + \frac{C}{b}$

C - Equilibrium concentration (ppm)

$\frac{x}{m}$ - amount adsorbed per unit weight of the adsorbent (soil)
b - adsorption maxima
k - Constant related to bonding energy

Q.36. Write the equation to calculate Lime potential of soils.

Ans. pH-l / 2p (Ca + Mg)

Q.37. Write Universal soil loss equation (USLE).

Ans. A = RKLSCP

A - Erosion soil loss (t acre-1 yr-1)
R - Rainfall factor
K - Soil erodibility factor
L - Slope length factor
S - Slope gradient factor
C - Vegetative cover
P - Practices used for erosion control

Q.38. What is the conversion factor for the conversion of Mg into MgO?

Ans. 1.658

Q.39. What is the conversion factor for the conversion of MgO into Mg?

Ans. 0.608

Q.40. What is the conversion factor for the conversion of Ca into CaO?

Ans. 0.399

Q.41. What is the conversion factor for the conversion of CaO into Ca?

Ans. 0.751

Annexure V
Differentiate

Q.1. Absorption vs Adsorption.

Ans. *Absorption* : Process by which a substance is taken into another substance.

Adsorption : It is a phenomenon of surface adherance. The attraction of ions or compounds to the surface of a solid.

Q.2. Active Absorption vs Passive Absorption.

Ans. *Active Absorption* : Process through which movement of ions and water into the plant root against an activity gradient. This process requires energy.

Passive Absorption : Transport of ions from the epidermis to the xylem without any consumption of metabolic energy.

Q.3. Adhesion vs Cohesion.

Ans. *Adhesion* : Process of molecular attraction that holds the surfaces of two dissimilar or unlike substances. Ex. Water and rock particles.

Cohesion : Process of attraction between like molecules

Q.4. Bulk density vs Particle density.

Ans. *Bulk density* : It is mass per unit volume of a dry soil (volume of solid + pore space).

Particle density : It is mass per unit volume of soil particles (soil solids) excluding porespace.

Q.5. State the difference between heat of solution and heat of wetting.

Ans. *Heat of Solution* is the energy released when ions become hydrated.

Heat of Wetting is the energy released when clay particles are hydrated.

Q.6. Active sensor vs Passive sensor.

Ans. *Active sensor* : The sensor produce its own electro magnetic radiations. Eg. RADAR

Passive sensor : which sense natural radiations either emitted or reflected from the earth. Eg. SUN

Q.7. Mass flow vs Diffusion.

Ans. *Mass flow* : Movement of nutrient ions and dissolved substances in the water during transpiration, water up take by the plant.

Diffusion : Movement of nutrient ions from a area of higher concentration to lower concentration.

Q.8. Biological Oxygen Demand (BOD) vs Chemical Oxygen Demand (COD).

Ans. BOD : Measure of the amount of O_2 required from water for biological decomposition of organic material.

COD : Measure of the O_2 consuming capacity of inorganic and organic matter present in water.

Q.9. *Nitrification Vs Denitrification.*

Ans. *Nitrification* : Process of biological oxidation by which NH_4-N is converted to NO_3-N.

Denitrification : Process of biochemical reduction of nitrate or nitrite to gaseous N, either as molecular N or as oxides of N.

Q.10. Endodynamorphic soils Vs Ectodynamorphic soils.

Ans. *Endodynamorphic soils* : Soils with properties that have been influenced by parent material.

Ectodynamorphic soils : Soils with properties that have been influenced by factors other than parent material.

Q.11. Pedology vs Edaphology.

Ans. *Pedology* : Branch of Soil Science that deals with survey, genesis and classification of soils for land use planning.

Edaphology : Branch of soil science that deals with the study of soils in relation to growth, nutrition and yield of crops.

Q.12. Eluviation Vs Illuviation.

Ans. *Eluviation* : Process of removal of soil materials from a horizon in solution or in suspension by percolating water when there is an excess of rainfall over evaporation

Illuviation : Process of accumulation of soil materials from an overlying layer and precipitated from solution or deposited from suspension.

Q.13. Mineralization vs Immobilization.

Ans. *Mineralization* : Conversion of an element from an organic form to an inorganic from as a result of microbial decomposition.

Immobilization : Conversion of an element from inorganic to organic form in microbial tissues or in plant tissues rendering the element unavailable to the plants.

Q.15. Extrusive rocks vs Intrusive rocks.

Ans. *Extrusive rocks (Volcanic)* : The rocks formed on the surface of earth where the molten material consolidates on cooling are called extrusive rocks.

Intrusive rocks : Rocks formed by the slow cooling of the molten magma within the earth surface.

Q.16. Hidden hunger of crops Vs Luxury consumption of nutrients.

Ans. *Hidden hunger of crops* : Condition in which crops need more nutrients yet does not show deficiency symptoms.

Luxury consumption of nutrients : Refers to the amount of nutrients absorbed by plants in excess of their requirement for optimum growth.

Q.17. Litho sequence Vs Catenary sequence.

Ans. *Lithosequence* : A group of related soils that differ from one another in certain soil characteristics as a result of variations in parent material.

Catenary sequence : A sequence of soils within a soil zone derived from similar parent material, climate and having different soil characteristics due to variations in relief and drainage.

Q.18. Petro genesis Vs Petrography.

Ans. *Petro genesis* : Study of the origin of rocks.

Petrography : Study of the description of rocks.

Q.19. pH dependent charge Vs Permanent charge.

Ans. *pH dependent charge* : The Al-OH or Si-OH portion of the clay ionizes the H and leaves a temporary negative charge on the oxygen (Al-O- or Si-O). Extent of ionized hydrogen depends on solution pH. Hence the charge on the clay surface depends on pH called pH dependent charge.

Permanent charge : Charges arise from the isomorphous substitution and not dependent on pH. Such charge on the clay surface is called permanent charge.

Q.20. Dispersion Vs Flocculation.

Ans. *Dispersion* : The colloidal particles tend to repel each other resulting in dispersion. Increase in Zeta potential cause dispersion or deflocculation

Flocculation : The colloidal particles tend to attract each other and coagulate like a bigger particle. Decrease in zeta potential cause flocculation.

Q.21. Active acidity Vs Exchange or Reserve acidity.

Ans. *Active acidity* : Acidity develops due to H^+ and Al^{3+} ions concentration of the soil solution. Magnitude of this acidity is low.

Exchange or Reserve acidity : Acidity developed due to adsorbed H^+ and Al^{3+} ions on the soil colloids. The magnitude of this acidity is very high.

Q.22. Laterisation Vs Podzolisation.

Ans. *Laterisation* : Process that removes silica and accumulate sesquioxides in the solum.

Podzolisation : Process that removes Ca^{2+} and other bases, sesquioxides from the surface horizon.

Q.22. pH Vs pE.

Ans. pH is negative logarithm of H+ ion activity

$$pH = \log\left[\frac{1}{H^+}\right]$$

pE is negative logarithm of electron activity

$$pE = \frac{Eh}{0.0591}$$

Q.23. Thermal conductivity Vs thermal diffusivity.

Ans. *Thermal Conductivity* : The change of heat dQ / dt flowing into or out of the soil dt / dZ in the soil.

$$\frac{dQ}{dt} = -K\frac{dt}{dz}$$

where K is thermal conductivity expressed in Joules/°C/cm /sec.

Thermal Diffusivity : Thermal conductivity (K) divided by the product of density of soil (P) and specific heat of the soil expressed in sq.cm per sec.

$$\text{Thermal diffusivity} = \frac{\text{Thermal conductivity}}{\text{Density of soil}} \times \text{specific heat of soil}$$

Q.24. Specific heat vs Heat capacity.

Ans. *Specific heat* : Amount of heat required to raise the temperature of one gram of a substance by 10°C. It is expressed in calorie per gm.

Heat capacity : Heat capacity of a given material (C) is equal to its specific heat (s) multiplied by the mass (m).

Heat capacity (Q) = s x m

Q.25. Hard pan Vs Duripan.

Ans. *Hard pan* : It is a hardened soil layer in the lower A or B horizon caused by cementation of soil particles with organic matter, silica, sequioxide and $CaCO_3$ etc. Hardness does not change with variation in soil moisture content.

Duripan : A sub surface horizon that is cemented by silica. Duripans can be disintegrated in acid or basic solutions.

Q.27. Leaching requirement vs Gypsum requirement.

Ans. *Leaching requirement (LR)* : Fraction of the irrigation water that must be leached through the root zone to control soil salinity at any specified level.

$$\frac{\text{LR Ddw}}{\text{Diw}} \times 100 = \frac{\text{Ec iw}}{\text{Ec dw}} \times 100$$

Ddw = Depth of drainage water (inches)
Diw = Depth of irrigation water (inches)
Ec iw = Electrical conductivity of irrigation water (dSm-1)
Ec dw = Electrical conductivity of the drainage water (dSm-1)

Gypsum requirement (GR) : Calculated amount of gypsum required to be added for reclamation of the alkali soil.

$$\text{GR (me of Ca}^{2+} \text{ / 100g soil)} = \frac{\text{ESP (initial) - ESP (Final)}}{100} \times \text{CEC}$$

Q.28. Deficiency vs toxicity of nutrients.

Ans. An element is deficient if its concentration is low enough to limit the crop growth severely and produces characteristic deficiency symptoms.

An element is toxic if the concentration is high enough to severely reduce the plant growth and yield.

Q.29. Differentiate chemical and nuclear reactions.

Ans. *Chemical reaction* : Interaction between atoms with energy changes, it does not affect an atomic nucleus, only rearrangement of valency electrons take place.

Nuclear reaction : Interaction between nucleus or nucleons with energy changes. It is a change in the number of neutrons or protons or both in atomic nucleus resulting in the formation of atomic nucleus of a new element.

Q.30. Differentiate nuclear fission and fusion reactions.

Ans. *Nuclear fission* : When a heavy nucleus is bombarded with neutrons, nucleus attain a state of excitation and split into two equal masses. It is self sustaining process.

Nuclear Fusion : Isotopes of very light elements such as $^1H_2$, 1H3 and 3Li6 react with one another to produce more stable heavier nuclei. It is also called as thermonuclear reaction.

Q.31. Manure vs Fertilizer.

Ans. Manures are organic in origin, capable of supplying plant nutrients upon decomposition and have low nutrient content. Fertilizers are inorganic capable of supplying plant nutrients in readily available form and has definite chemical composition and high analytical value.

Annexure VI

Scientists and Innovations

	Name of the Scientist	Contribution
1.	Henri Becquerel	Natural radioactivity
2.	Madame Curie	Artificial radioactivity
3.	Aston, F.W,	Transmission of elements
4.	Soddy	Group displacement law
5.	Hillel	Father of soil physics
6.	Jenny	Age of soil, Oscillation volume
7.	Myers	Five stages of soil development
8.	Vilenski	Soil classification based on the climate
9.	Marbut	Soil groups, Pedocal, Pedalfer, Morpho genetic - system of soil classification, Soil classification based on zonality
10.	Dokuchaiev	Father of soil science, Founder of Modern Pedology, Genetic System of Soil Classification
		Soil Classification based on soil properties
11.	Mohr	Weathering stages in soil formation
12.	Viswanath & Ukil	First soil map of India
13.	Joffe	Defined soil as dynamic natural body Soil classification based on soil properties
14.	Hilgard	Defined soil as Loose and friable material
15.	Jackson	Weathering sequence of minerals
16.	Martin and Russel	Y-index
17.	Klingebiel and Montgomery (1961)	First land capability classification
18.	Guldberg and Waage	Law of mass action
19.	Debye-Huckel	Ion activity Co-efficient
20.	Lewis and Randall	Concept of ionic strength
21.	Waksman (1932)	Lignin theory of humus formation
22.	Stevenson (1982)	Polyphenol theory of humus formation, Sugar-amine condensation theory of humus formation
23.	Thomas, J.	Father of soil chemistry

24.	Justus Von Liebig (1840)	Law of minimum. Father of agricultural chemistry
25.	Boussingault	Father of Field experiment
26.	Nicholas	Functional metabolism of nutrients
27.	Mitscherlich (1909)	Law of Diminishing Return, Nutrient diagnosis by pot culture method, Law of physiological relationship, growth law
28.	Bray (1909)	Nutrient Mobility Concept
29.	Mehlich	Nutrient diagnosis in soil by Aspergillus niger method
30.	Sackett and Stewart	Nutrient diagnosis in soil by plaque method
31.	Cate and Nelson(1965)	Critical soil test level approach
32.	John B Laws	First to manufacture super phosphate
33.	Arnon and stout(1939)	Essentiality of nutrients, Essentialityof Mo to plants
34.	Gris (1843)	Essentiality of Fe to plant
35.	Theodore de Saussure (1804)	Essentiality of N to plants
36.	Warrington, K. (1923)	Essentiality of B to plants
37.	Sommer and Lipman (1926)	Essentiality of Zn to plants
38.	Page and Haver	Lattice hole theory
39.	Munsell	Soil colour chart
40.	Atterberg	Soil moisture limit
41.	Priestley (1800)	Essentiality of carbon to plants
42.	Carl sprengel (1839)	Essentiality of P, K, Ca, Mg to plants
43.	Sachs and Knop (1860)	Essentiality of S to plants
44.	Broyer, Carlton, Johnson, Stout (1954)	Essentiality of Cl to plants
45.	Mc Hargue (1922)	Essentiality of Mn to plants
46.	Sommer, Lipman and Mc Kinney (1931)	Essentiality of Cu to plants
47.	Brown welch, Cary (1987)	Essentiality of Ni to plants
48.	Clements (1961)	Crop logging in sugarcane
49.	Parker (1951)	Nutrient indexing

50.	Fried and Dean (1954)	'A' value technique
51.	Larsen (1952)	'L' value technique
52.	Russel (1954)	'E' value technique
53.	Buol (1975)	Fertility Capability Classification
54.	Colwell (1967)	Deductive approach
55.	Ramamoorthy (1968)	Inductive approach
56.	Beaufils	DRIS index
57.	Mulder	Nutrient interaction
58.	Hutchinson & Richards	ADCO composting process
59.	Howard and ward	Indore composting process
60.	Acharya C.N.	Bangalore method of composting

❑❑❑

MODEL QUESTION PAPERS

Model Question Paper - I

Q.1. 'A' horizon of the soil is also designated as :

a) Eluviated horizon
b) Illuviated horizon
c) Both a & b
d) None of these

Q.2. 'Hue" is the term which is used for the determination of :

a) Relative thickness or whiteness
b) Purity of the colour
c) Dominant spectral or rainbow
d) All of the above

Q.3. A combination of many taxonomic unit like soil series, phase or types is called :

a) Soil variant
b) Soil series
c) Soil complex
d) Soil association

Q.4. A man made epipedon which is similar to mollic epipedon in many characters except P_2O_5 content (more than 250 ppm):

a) Histic epipedon
b) Ochric epipedon
c) Anthropic epipedon
d) Umbric epipedon

Q.5. A sequence of soil of about same age, derived from same parent material and occurring under similar climatic conditions but have different characteristics due to variation in relief and drainage, is termed as :

a) Soil series b) Catena
c) Soil phase d) None of the above

Q.6. A surface horizon which has very high organic matter is :

a) Histic epipedon b) Ochric epipedon
c) Umbric epipedon d) None of the above

Q.7. A thick dark coloured horizon which have high base saturation (50% or more), high organic matter content and with strong soil structure, is called :

a) Umbric epipedon b) Mollic epiedon
c) Ochric epipedon d) Histic epipedon

Q.8. A unit of soil classification as well as soil mapping which is similar to soil series in most of the soil properties except one or two but the area of such a unit is not enough to establish it as a separate series and show on a soil map, is known as :

a) Soil association b) Soil phase
c) Soil variant d) None of the above

Q.9. A vertical section of soil body which shows different layers is known as :

a) Soil horizons b) Soil profile
c) Regolith d) Parent material

Q.10. A vertical section of the soil through all its horizons are extending into the parent material is known as :

a) Soil horizons b) Solum
c) Profile d) None of the above

Q.11. A well developed soil have :

a) A horizon b) B horizon
c) C horizon d) All of the above

Q.12. According to land capability classification, the soils which are not suitable for crop cultivation belong to :

a) Class I
b) Class III
c) Class IV
d) Class VIII

Q.13. Active soil forming factors would include :

a) Vegetation and topography
b) Time, topography and climate
c) Time and vegetation
d) Vegetation and climate

Q.14. Aeolian deposits are formed by the transportation of materials through :

a) Running water
b) Wind
c) Sea
d) Ice

Q.15. All India soil survey and land use organization was established in :

a) 1952
b) 1956
c) 1960
d) 1962

Q.16. Amount of K in feldspar and mica is about :

a) 60%
b) 20%
c) 90%
d) 10%

Q.17. An abrupt change in textural class or no continuity found between B and C horizon is termed as :

a) Textural change
b) Lithological discontinuity
c) Regolith
d) None of the above

Q.18. An argillic sub-surface horizon with high E.S.P. (more than 15) is called :

a) Cambic horizon
b) Nitric horizon
c) Agric horizon
d) Oxic horizon

Q.19. An epipedon which has dark colour and low in base saturation is called :

a) Umbric epipedon
b) Histic epipedon
c) Ochric epipedon
d) Anthropic epipedon

Q.20. Apatite is a group of minerals which mainly contains :

a) Sulphur
b) P_2O_5
c) N_2
d) Titanium

Q.21. Two elements that make up more than 70% of the earth crust:

a) Silicon and Aluminium
b) Oxygen and Silicon
c) Oxygen and Aluminium
d) Silicon and Iron

Q.22. A compound that occurs in more than one crystal form:

a) Mineral
b) Mineraloid
c) Polymorph
d) Cryptocrystalline

Q.23. How many grams of $CaCO_3$ per 100 grams of soil would be required to neutralize 6 meq.of acid? (M.W. of $CaCO_3$ = 100, Valency is 2):

a) 30
b) 0.30
c) 3.0
d) 0.03

Q.24. Bonding occurs between ions of opposite charge:

a) Electrostatic bonding
b) Co-valent bonding
c) Hydrogen bonding
d) Vanderwaals force

Q.25. The arrangement of ions or atoms that is repeated at regular intervals in three dimension is called:

a) Crystal
b) Unit cell
c) Basal plane
d) Radius ratio

Q.26. Chlorite is a

a) 2:1 type mineral
b) 1:1 type mineral
c) 2:1:1 type mineral
d) 1:2 type mineral

Q.27. Clay minerals can be identified and quantified by:

a) Mass spectrometry
b) X ray diffraction method
c) Turbidimetry
d) Gravimetric method

Q.28. Black soils have high content of :

a) R_2O_3
b) Al_2O_3
c) Fe_2O_3
d) Silica

Q.29. X/m = KC n is :

a) Langmuir equation
b) Freundlich's equation
c) Gapon 's equation
d) Vanselow 's equation

Q.30. Buffering capacity is high for the soils with :

a) 1:1 clay mineral
b) 2:1 clay mineral
c) 2:1:1 clay mineral
d) All

Q.31. The CEC of vermiculite is:

a) 15 meq/100g
b) 30 meq/100g
c) 20-40 meq/100g
d) 150 meq/100g

Q.32. Which one of the following is responsible for the dispersion of soil particles?

a) Exchangeable calcium
b) Exchangeable magnesium
c) Exchangeable sodium
d) Exchangeable potassium

Q.33. The pH was given by:

a) Sorensen
b) Schofield
c) White and Beckett
d) Thomas Way

Q.34. The emf of a cell can be measured by :

a) Hydrogen electrode
b) Reference electrode
c) Calomel electrode
d) Indicator electrode

Q.35. Release of K+ ions from mica leads to the formation of :

a) Illite
b) Vermiculite
c) Montmorillonite
d) Chlorite

Q.36. Under normal field conditions, the soil water potential is:

a) Positive
b) Negative
c) Zero
d) Unity

Q.37. Clay minerals are:

a) Lyophyllic colloids
b) Lyophobic colloids
c) Hydrophillic colloids
d) Hydrophobic colloids

Q.38. What is the conversion factor for the conversion of nitrogen into protein?

a) 6.25 b) 6.724

c) 6.75 d) 6.50

Q.39. Of the following, which theory is also called as diffuse-double layer theory?

a) Helmholtz double layer theory

b) Guoy-Chapman double layer theory

c) Stern double layer theory

d) None

Q.40. Adding organic matter to a soil would

a) Increase the CEC

b) Increase the water holding capacity

c) Both of the above

d) None of the above

Model Question Paper - II

Q.1. The permissible maximum biuret content of urea produced in India is

a) 0.5% b) 1.0%
c) 1.5% d) 2.0%

Q.2. The ion that is more susceptible to leaching is

a) NH_4^+ b) NO_3^-
c) $H_2PO_4^-$ d) HPO_4^{2-}

Q.3. The predominant ionic species of phosphate in soil at pH below 6.7 is

a) PO_4^{3-} b) HPO_4^{2-}
c) $H_2PO_4^-$ d) PO_4^{2-}

Q.4. Which fertilizer contains nutrients other than primary nutrients?

a) Diammonium phosphate b) Urea
c) Ammonium sulphate d) Ammonium nitrate

Q.5. In anaerobic soils, sulphate is reduced by

a) Thiobacillus b) Desulfotomaculum
c) Nitrosomonas d) Nitrococcus

Q.6. Father of the field-plot method of experimentation

a) Jean Baptisete Boussingault b) Liebig
c) Bray d) Steinberg

Q.7. Succulence of the plants increases due to excessive consumption of

a) Nitrogen b) Calcium
c) Boron d) Potassium

Q.8. .... imparts disease resistance to crops

a) Potassium b) Nitrogen
c) Hydrogen d) Chlorine

Q.9. Chloride-loving crops

a) Coconuts b) Pulses
c) Sorghum d) Rice

Q.10. Blossom end rot in tomatoes is due to – – – deficiency

a) N b) B
c) Ca d) S

Q.11. Bronzing disease in rice is due to Toxicity of – – – in soil

a) Cu b) B
c) Zn d) Fe

Q.12. Involved in the splitting of H_2O molecules during photosynthesis

a) B b) Cu
c) Mn d) Fe

Q.13. – deficiency in cauliflower is termed as whip-tail.

a) S b) N
c) Mn d) Mo

Q.14. Hen and chick disease in grapes is due to – – deficiency

a) H b) Cu
c) B d) S

Q.15. "Criteria of essentiality" proposed by

a) Liebig b) Bray
c) Baule d) Arnon and Stout

Q.16. Essential nature of N first established by

a) Baule b) Theodre de Saussure
c) Arnon and Stout d) Liebig

Q.17. Site for Ion exchange

a) Clay b) Silt
c) Organic matter d) All

Q.18. – – is the dominant exchangeable ion in the soil

a) Ca^{2+} b) H^+
c) K^+ d) Na^+

Q.19. The sum of the milliequivalents of calcium, magnesium, potassium and sodium plus aluminium and hydrogen is the

a) CEC b) BSP
c) Effective CEC d) AEC

Q.20. Luxury consumption is related with the element

a) P b) K
c) N d) Ca

Q.21. Which method is used for the determination of dissolved oxygen?

a) Winuler method b) Walkley & Black method
c) Bray and Kurtz method d) All

Q.22. Which of the following is an alkalinity end indicator?

a) Phenolphthalein b) Methyl orange
c) Thymal blue d) All

Q.23. Total hardness of water sample contains only

a) Ca^{2+} b) Mg^{2+}
c) K^+ d) Both a and b

Q.24. Sodium can be determined by

a) Gravimetric method b) Volumetric method
c) Flame photometer method d) All

Q.25. Which of the following is a soil dispersing agent?

a) Na fluoride b) Na chloride
c) Na phosphate d) Na hexa meta phosphate

Q.26. Which of the following reducing agent is required for the removal of free iron oxides?

a) Sodium dithionate b) Sodium citrate
c) Sodium fluoride d) A and B

Q.27. What is the role of calcium in mono calcium phosphate used as an extractant in available S estimation?

a) Increase flocculation
b) Decrease flocculation
c) Prevent S extraction from organic S
d) Improve S extraction from organic S

Q.28. What is the colour formed by the reaction of iron with orthophenon throline in colorimetric determination of iron?

a) Green b) Blue
c) Red d) Yellow

Q.29. What is the colour formed by the reaction of Mn with potassium periodate in strongly acid medium in colorimetric determination of Mn

a) Pink b) Blue
c) Red d) Yellow

Q.30. RSC index for safe use of water concept was given by

a) Eaton b) Wilcox
c) Arnon d) Stout

Q.31. Which is the detector of ICP-OES?

a) Photo multiplier tube b) Solid state detector
c) Both A and B d) None of the above

Q.32. Which of the following can effectively absorb impurities in water samples

a) Dargo G -60 b) Ethyl alcohol
c) Gum acacia d) Agar

Q.33. Which of the following is nessler reagent?

a) I_4HgK_2 b) $HgOHg(NH_2)I$
c) KOH d) None

Q.34. In total digestion for nitrogen estimation, which prevents loss of nitrate nitrogen?

a) Salicylic acid b) Malic acid
c) Oxalic acid d) Formic acid

Q.35. Which of the following is the type of detector used in HPLC?

a) Absorbants detector
b) Electrochemical detector
c) Conductivity detector
d) All the above

Q.36. The electrode of pH meter consist of element in contact with a solution of KCl

a) Hg
b) Mg
c) Zn
d) Ag

Q.37. Quantitative law basic to colorimetric analysis is ....

a) Beer's law
b) Lambert's law
c) A and B
d) None

Q.38. The inside of the bulb in a pH meter is filled with an electrolyte of high buffer capacity which is of

a) 0.1 N HCl
b) 1 N HCl
c) 0.5 N HCl
d) 0.05 N HCl

Q.39. Principle of operation in AAS is

a) Sample aerosal is aspirated
b) Reduction of element to atomic state
c) Absorption of radiation directly proportional to number of atoms
d) All

Q.40. The ICP works on

a) emission spectrometry
b) Optical emission spectrometry
c) Absorption spectrometry
d) All

Q.41. What is the number of classifications in size groups of soil particles by International Society of Soil Science

a) 3
b) 4
c) 6
d) 7

Q.42. The equation $v=Kr^2$ is called as

a) Stoke's law
b) Ratio law
c) Leobig law
d) Mitcherlich equation

Q.43. More accurate analysis of soil texture can be done by

a) Hydrometer method
b) International pippette method
c) Aggregate method
d) All

Q.44. What is the weight of a hectare of surface soil

a) 2.25×10^6 kg ha^{-1}
b) 2×10^6 kg ha^{-1}
c) 2.4×10^6 kg ha^{-1}
d) 2.6×10^6 kg ha^{-1}

Q.45. In puddle soil the soil particles are ....

a) Cohesive
b) Massive
c) Non- cohesive
d) Both A and B

Q.46. The method of soil structure characterization and evaluation that was described by Yodder is

a) Stability index
b) Wet sieving
c) Dry sieving
d) MWD

Q.47. The physical condition of soil in relation to plant growth is called as

a) Soil tilth
b) Tillage
c) Coherance
d) Adherance

Q.48. The distance travelled by water through soil column is called as

a) Infiltration rate
b) Percolation rate
c) Seepage rate
d) Run off loss

Q.49. The formula $q=k \Delta H/L$ is

a) Darcy's law
b) Hydraulic gradient
c) Flux density
d) Fick's law

Q.50. In the field, hydraulic conductivity is measured by

a) Piezometer
b) Constant Head permeameter
c) Auger hole method
d) A and B

Q.51. What is the unit of surface tension?

a) Dynes cm^{-1}
b) cm sec^{-1}
c) g cm^{-3}
d) Poise

Q.52. What is the soil water potential at field capacity ?

a) 10-33 kPa
b) 33 kPa
c) 1500 kPa
d) 33-1500 kPa

Q.53. In-situ measurement of soil water can be done by

a) Electrical resistance method
b) Neutron scattering method
c) Gamma rays attenuation technique
d) All

Q.54. Diffusion of gases under total pressure gradient takes place by

a) Mass flow
b) Diffusion
c) Partial diffusion
d) All

Q.55. What is the unit of specific heat?

a) kcal g^{-1}
b) kcal cm^{-3}
c) cal cm^{-3}
d) cal g^{-1}

Q.56. The dominant spectral colour in Munsell colour chart indicates

a) Hue
b) Value
c) Chroma
d) All

Q.57. What is the unit of Oxygen Diffusion Rate?

a) $g/cm^2/min$
b) cm/g/min
c) min /cm / g
d) None

Q.58. Bulk density of organic matter is

a) 0.3 g/cc
b) 0.5 g/cc
c) 0.6 g/cc
d) 1.0 g/cc

Q.59. What is the critical limit of Oxygen Diffusion Rate?

a) 20×10^{-8} g/cm^2/min
b) 22.9×10^{-8} g/cm^2/min
c) 3.9×10^{-8} g/cm^2/min
d) 25×10^{-8} g/cm^2/min

Q.60. Soil compaction increases the

a) Bulk density
b) Hydraulic conductivity
c) Macropores
d) Infiltration rate

Model Question Paper - III

Q.1. The dominant soil group in India is.........

a) Red soil b) Black soil
c) Alluvial soil d) Laterite soil

Q.2. Diagnostic surface horizon is termed as ....

a) Endopedon b) Pedon
c) Control section d) Epipedon

Q.3. The standard colour used to indicate class II of LCC is ....

a) Green b) Dark green
c) Orange d) Yellow

Q.4. Pesticide consumption is highest in............

a) Rice b) Wheat
c) Cotton d) Vegetables

Q.5. Which one of the following is a metamorphic rock?

a) Granite b) Basalt
c) Marble d) Sandstone

Q.6. The NBSS & LUP is located at.......

a) Delhi b) Nagpur
c) Calcutta d) Hyderabad

Q.7. The type of soil survey employed for micro level planning is ....

a) Exploratory b) RSS
c) Rapid RSS d) DSS

Q.8. Which one of the following is grouped under non-silicate mineral ?

a) Hematite b) Chlorite
c) Vermiculite d) Illite

Q.9. According to "Soil Taxonomy" ....number of soil orders so far recognized?

a) Ten b) Eleven
c) Twelve d) Twenty

Q.10. The concept of pH was introduced by..........

a) Sorenson b) Schofield
c) Waksman d) Richards

Q.11. The ion that causes most rapid deterioration in soil structure is ....

a) Mg b) Ca
c) Na d) H

Q.12. The soil forming process responsible for formation of Spodosol is ....

a) Laterisation b) Pedoturbation
c) Calcification d) Podzolization

Q.13. The land capability class not having any sub class is ....

a) Class VIII b) Class V
c) Class I d) Class II

Q.14. The most common spectrum of electromagnetic radiation used for natural resource mapping is ....

a) Visible b) UV
c) IR d) Microwaves

Q.15. The satellite placed 36,000 kms away from the earth is termed as............

a) Polar orbiting b) Sun synchronous
c) Equator orbiting d) Geo-stationary

Q.16. The NRSC is located at ....

a) Nagpur b) Hyderabad
c) Chennai d) Delhi

Q.17. The term "gilgai micro relief" is associated with .... soil order

a) Vertisols b) Oxisols
c) Histosols d) Gelisols

Q.18. The formative element of Inceptisols is ....

a) inc b) nce
c) cep d) ept

Q.19. ....is the Indian satellite used for collecting meteorological data

a) INSAT 2E b) IRS P_4
c) IRS ID d) SPOT

Q.20. Forest plantations of an area is observed as.............colour in FCC

a) Red b) Green
c) Blue d) Black

Q.21. Salt affected areas are observed as.........colour in FCC

a) Black b) Yellow
c) Dark blue d) White-grayish

Q.22. Salinity of irrigation water alters...........of the soil pollution

a) Matric potential b) Osmotic potential
c) Redox potential d) Electrical potential

Q.23. ....is used for reclaiming acid soil

a) Gypsum b) Sulphur
c) $FeSO_4$ d) Lime

Q.24. Sub soil hard pan is mostly observed in ....

a) Histosols b) Gelisols
c) Alfisols d) Vertisols

Q.25. The problem of surface crusting is due to presence of .... ions

a) Ca & Mg b) Mg & K
c) Fe & Al d) Na & K

Q.26. Central Soil Salinity Research Institute is located at ....

a) Delhi b) Karnal
c) Trichy d) Ludhiana

Q.27. The specific surface area is very high in ....

a) Sand b) Silt
c) Clay D) Gravel

Q.28. ....crop is highly tolerant to soil salinity

a) Rice b) Ragi
c) Black gram d) Bengal gram

Q.29. Electrical conductivity of soil is a measure of...........

a) Hydrogen b) Salts
c) Carbon d) Negative charges

Q.30. Conversion of organic form of N into inorganic form is known as.........

a) Mineralization b) Immobilization
c) Fixation d) Synthesis

Q.31. The phytotoxic substance present in urea as an associated impurity is

a) Bisect b) Biform
c) Biuret d) Urease

Q.32. The form of N present in urea is ....

a) Ammonical b) Amine
c) Amide d) Nitrate

Q.33. Name the nutrient which is taken up by plants both in anionic and cationic form

a) Potassium b) Calcium
c) Zinc d) Nitrogen

Q.34. The most suitable P fertilizer for acid soil is ....

a) Super phosphate b) Rock phosphate
c) Triple super d) DAP

Q.35. Khaira disease in rice is caused due to the deficiency of ....

a) Mo b) B
c) Cu d) Zn

Q.36. The N content of ammonium sulphate fertilizer is.............

a) 20.6% b) 22.6%
c) 26.0% d) 28.0%

Q.37. The "inductive approach" of soil fertility evaluation was proposed by..............

a) Cowell b) Ramamoorthy et al
c) N.S.Randawa d) Cate & Nelson

Q.38. .... elements forms structural component of chlorophyll

a) K b) Na
c) Mg d) P

Q.39. "Cynamide process" is employed in manufacture of .... fertilizers

a) K b) Zn
c) Mo d) N

Q.40. The downward movement of water in soil is termed as ....

a) Seepage b) Run-off
c) Infiltration d) Percolation

Q.41. The AEC of the clay mineral depends upon it's................

a) Particle size b) Structure
c) Amount of positive charges d) None of these

Q.42. Poorly aerated soils are significant sources of

a) Methane gas b) Ammonia gas
c) Nitrogen gas d) Argon gas

Q.43. Water is not available to plants but may be available to certain bacteria at atmosphere tension of..................

a) 31 bars
b) 1000bars
c) 10000 bars
d) 100 bars

Q.44. Except one nutrient, all others are naturally available in soil through clay minerals

a) Calcium
b) Nitrogen
c) Magnesium
d) Aluminum

Q.45. The relationship between CEC and AEC is..........

a) Inverse
b) Synergistic
c) Both A & B
d) None of these

Q.46. The pulse protein is dominated by amino acid..................

a) Cystein
b) Methionine
c) Lysine
d) Glysine

Q.47. Synthetic pyrtheriods are widely accepted for use because of it's ....

a) Rapid action
b) Low mammalian toxicity
c) Lack of persistence
d) Selective toxicity

Q.48. Mohr's scale is used to measure..............of the mineral

a) Lusture
b) Colour
c) Hardness
d) Specific gravity

Q.49. The fruit rich in fat.........

a) Avocado
b) Grapes
c) Apple
d) Papaya

Q.50. The awful odour of coconut oil during deterioration is due to ....

a) d-Lactone
b) n-Octanol
c) Methyl ketone
d) Tocopherols

Q.51. Which one of the following is a communication satellite ....

a) Rohini
b) Aryabatta
c) Apple
d) Baskara

Q.52. The enzyme which catalyses the reduction of hydroxyl amine to NH_3 requires the co-factor.........

a) Mn b) Mo
c) Zn d) Fe

Q.53. Hydroxy unsaturated fatty acid present in castor..........

a) Oleic acid b) Linoleic acid
c) Ricinoleic acid d) Linolenic acid

Q.54. The flavouring agent in banana.......

a) Amyl acetate b) Terpenes
c) Methyl anthranilated d) Ethyl 2 –methyl butrate

Q.55. An example for neutral nitrogenous fertilizer is ....

a) Urea b) Ammonium chloride
c) Ammonium sulphate d) CAN

Q.56. The cyanogenic glucoside in young sorghum plants...........

a) Dhurrin b) Sinigrin
c) Arbutin d) Baptisin

Q.57. Rotenone is the principle toxin in ....

a) Derris b) Chrysanthemum
c) Tobacco d) Neem

Q.58. Endosulfan belongs to ....

a) Synthetic pyrethroids b) Organo phosphorus
c) Cyclodiene d) Carbamate

Q.59. Enzyme derived from pine apple juice used for protein digestion...............

a) Papain b) Chymopapain
c) Bromealin d) None of the above

Q.60. The metal ion present in Zineb is ....

a) Al b) Zn
c) Hg d) Fe

❑❑❑

About the Authors

DR. R.K. KALEESWARI, working as Professor (Soil Science) has 17 years of teaching and research experiences. She has completed her post doctoral research programme in Australia to explore the possibilities of soil carbon sequestration in salt affected soils. Her field of specialization is soil fertility with regard to soil carbon sequestration in agro-ecosystems. She has operated 10 research projects on soil fertility. She is awarded with prestigious award of Government of Australia, Endeavour Research award-2008.Tamil Nadu Agricultural University awarded her with Best Researcher award during the year 2009.

DR. M.R. LATHA working as Assistant Professor (Soil Science) has 8 years experience in teaching, research and extension activities. In her career spanning over eight years, she has published 29 research papers in reputed Journals.

DR. R. RAJESWARI working in Soil Science and Agricultural Chemistry, Tamil Nadu Agricultural University, Coimbatore. She got Dr.K.K.Krishnamoorthy and Dr.G.Selvakumari awards in TNAU. She has six years of professional experience. She has published 6 National 4 International Papers and 10 Abstracts.